Springer Undergraduate Mathematics Series

Editor-in-Chief
Endre Süli, Oxford, UK

Series Editors
Mark A. J. Chaplain, St. Andrews, UK
Angus Macintyre, Edinburgh, UK
Shahn Majid, London, UK
Nicole Snashall, Leicester, UK
Michael R. Tehranchi, Cambridge, UK

The Springer Undergraduate Mathematics Series (SUMS) is a series designed for undergraduates in mathematics and the sciences worldwide. From core foundational material to final year topics, SUMS books take a fresh and modern approach. Textual explanations are supported by a wealth of examples, problems and fully-worked solutions, with particular attention paid to universal areas of difficulty. These practical and concise texts are designed for a one- or two-semester course but the self-study approach makes them ideal for independent use.

Roozbeh Hazrat

A Course in Python

The Core of the Language

 Springer

Roozbeh Hazrat
Centre for Research in Mathematics and Data Science
Western Sydney University
Penrith, NSW, Australia

ISSN 1615-2085 ISSN 2197-4144 (electronic)
Springer Undergraduate Mathematics Series
ISBN 978-3-031-49779-7 ISBN 978-3-031-49780-3 (eBook)
https://doi.org/10.1007/978-3-031-49780-3

Mathematics Subject Classification (2020): 00-01

© The Editor(s) (if applicable) and The Author(s), under exclusive license to Springer Nature Switzerland
AG 2023
This work is subject to copyright. All rights are solely and exclusively licensed by the Publisher,
whether the whole or part of the material is concerned, specifically the rights of translation, reprinting,
reuse of illustrations, recitation, broadcasting, reproduction on microfilms or in any other physical way,
and transmission or information storage and retrieval, electronic adaptation, computer software, or by
similar or dissimilar methodology now known or hereafter developed.
The use of general descriptive names, registered names, trademarks, service marks, etc. in this publication
does not imply, even in the absence of a specific statement, that such names are exempt from the relevant
protective laws and regulations and therefore free for general use.
The publisher, the authors, and the editors are safe to assume that the advice and information in this
book are believed to be true and accurate at the date of publication. Neither the publisher nor the
authors or the editors give a warranty, expressed or implied, with respect to the material contained herein
or for any errors or omissions that may have been made. The publisher remains neutral with regard to
jurisdictional claims in published maps and institutional affiliations.

This Springer imprint is published by the registered company Springer Nature Switzerland AG
The registered company address is: Gewerbestrasse 11, 6330 Cham, Switzerland

Paper in this product is recyclable.

Preface

Python has become, for good reason, a very popular programming language, with a substantial number of followers around the globe. In addition to its intuitive programming language, Python now offers numerous libraries which provide powerful tools and methods to do, among many other things, mathematics, physics, text processing, and musical composition. Once one grasps the core of the language, then help, hints, and a wealth of sample codes are just a google away.

Besides many books on Python, there is an uncountable number of documents on the internet. However, I wanted to have one source that I can follow to systematically teach (or learn) the core of the language before diving into more advanced fronts. I also wanted to have a source that is short, to the point and which provides interesting programming examples; examples that I enjoy coding, modifying and experimenting with.

This book grew out of a course I gave at Western Sydney University. It allows the reader to learn Python by going through interesting exercises step by step, with short and concise explanations. I have tried to let the reader learn from the codes and avoid long and exhausting explanations, as the codes will speak for themselves. Also I have tried to inspire the reader's imagination by showing that in Python (as in the real world) there are many ways to approach a problem and solve it.

Thus, this book could be considered for a course in Python, or for self-study. It mainly concentrates on the core of Python programming. I have mostly chosen problems having something to do with natural numbers as they do not need any particular background. The codes have been written in Jupyter using Python version 3.

Acknowledgement. My thanks go to Thomas Fischbacher from Google Research for going through the book and providing insightful comments; to the students at Western Sydney University for their contributions to the classes, which made them lively and enjoyable, and to the anonymous referees for their support and feedback. My thanks also go to Dr. Remi Lodh at Springer for his generous support throughout the preparation of the book.

How to use the book. This book can be used as a one semester course in programming in Python (13 weeks, 3 hours each week) at the undergraduate level or a compact crash course (4 days, 6 hours each day) for more advanced students. Each chapter starts with a description of new tools and topics and provides examples. The examples are chosen so that the reader can learn from the codes by typing and running them and then modifying them and experimenting with them. Long and exhausting explanations are avoided. The lecturer can code the examples in the class along with students. The students are then encouraged to take the lead role and compose codes themselves for some of the exercises. The problems at the end of each chapter can be given to students as homework so that they present the codes in the class the next week. The book concludes with projects which demonstrate how to use Python to explore mathematics. Some of these projects can be given as assignments to the students.

Please start with the codes in the book, change them, tear them apart and turn them upside down, and create your own better programs.

Sydney, Australia, October 2023   *Roozbeh Hazrat*
                                                         r.hazrat@westernsydney.edu.au

Contents

Chapter 1

Basics of Python

*1.1 Basic Arithmetic

This first chapter gives an introduction to using Python "out of the box", demonstrating how to use ready-made commands, performing basic arithmetic and building up computations. One of the reasons why Python has become the language of choice for many is its external libraries, which provide powerful tools that one can use to investigate and analyse problems in a multitude of areas. We will showcase some of them in this first chapter, such as working with symbols and the first instances of symbolic computations.

Python handles all sort of numerical calculations, both exact computations and approximations. If we would like to calculate $3 \times 4 \times 5 \times 6 + 1$ or 2^{3^4} we only need to enter them correctly into Python.

```
[1]: 3 * 4 * 5 * 6 + 1
```

```
[1]: 361
```

```
[2]: 2**3**4
```

```
[2]: 2417851639229258349412352
```

Python is a powerful calculator with the basic arithmetic operations; +, − for addition and subtraction and ∗, ∗∗ for multiplication and for raising to powers. One uses round brackets () to group the expressions together.

```
[3]: (2**3)**4
```

```
[3]: 4096
```

```
[4]: 2**(3**4)
```

© The Author(s), under exclusive license to Springer Nature Switzerland AG 2023
R. Hazrat, *A Course in Python*, Springer Undergraduate Mathematics Series,
https://doi.org/10.1007/978-3-031-49780-3_1

`[4]:`   241785163922925834941235241785163922925834941235

Here is a calculation to disprove a 200-year-old conjecture of the famous mathematician Euler. He conjectured that three fourth powers can never sum to a fourth power.

```
[5]:  2682440**4 + 15365639**4 + 18796760**4
```

`[5]:`   180630077292169281088848499041

```
[6]:  20615673**4
```

`[6]:`   180630077292169281088848499041

The last two calculations show that

$$2682440^4 + 15365639^4 + 18796760^4 = 20615673^4,$$

which was found by Noam Elkies at Harvard in 1988, giving a counterexample to Euler's conjecture.

Python is even more powerful; it can handle large exact computations, as the following shows.

```
[7]:  2**9941 - 1
```

`[7]:`   34608828249085121524296039576741331672262866890023854779048928344500622080983411446436437554415370753366448674763505018641470709332373970608376690404229265789647993709760358469552319045484910050304149809818540283507159683562232941968059762281334544739720849260904855192770626054911793590389060795981163838721432994278763633095377438194844866471124967685798888172212033000821469684446495614699719412692128433620646331385953757720046244202906468132608755825748847048938424398927023688497864306309300442293960337001054659538630200907304394448220255909740670059733305707995078329631309387398850801984162586351945229130425629366798595874957210311737477964188950607019417175060019371524300323636319342657985162360474512090898647074307803622983070381934454864937566479918042587755749738339033157350828910293923593527586171850199425548346718610745487724398807296062449119400666801128238240958164582617618617466040348020564668231437182554927847793809917495802552633233265364577438941508489539699028185300578708762293298033382857354192282590221696026655322108347896020516865460114667379813060562447480055071718250333737502267307344178512950738594330684340802698228963986562732597175372087295649072830289749771358330867951508710859216743218522918811670637448496498549094430

```
412774440794079895398574694527721321665808857543604774088429133
272929486968974961416149197398454328358943244736013876096437505
146992150326837445270717186840918321709483693962800611845937461
435890688111902531018735953191561073191960711505984880700270887
058427496052030631941911669221061761576093672419481606259890321
279847480810753243826320939137964446657006013912783603230022674
342951943256072806612601193787194051514975551875492521342643946
459638539649133096977765333294018221580031828892780723686021289
827103066181151189641318936578454002968600124203913769646701839
835949541124845655973124607377987770920717067108245037074572201
550158995917662449577680068024829766739203929954101642247764456
712221498036579277084129255555428170455724308463899881299605192
273139872912009020608820607337620758922994736664058974270358117
868798756943150786544200556034696253093996539559323104664300391
464658054529650140400194238975526755347682486246319514314931881
709059725887801118502811905590736777711874328140886786742863021
082751492584771012964518336519797173751709005056736459646963553
313698192960002673895832892991267383457269803259989559975011766
642010428885460856994464428341952329487874884105957501974387863
531192042108558046924605825338329677194691145990192132498496880
100211899682849413315731640563047254808689218234425381995903838
524127868408334796114199701017929783556536507553291382986542462
253468272075036067407459569581273837487178259185274731649705820
951813129055192427102805730231455547936284990105092960558497123
779789849218399970374158976741548307086291454847245367245726224
501314799926816843104644494390222505048592508347618947888895525
278984009881962000148685756402331365091456281271913548582750839
078914699790194262248837894635551
```

If a number of the form $2^n - 1$ happens to be prime, it is called a *Mersenne prime*. Recall that a *prime number* is a number greater than 1 which is divisible only by 1 and itself. It is easy to see that $2^2 - 1$, $2^3 - 1$ and $2^5 - 1$ are Mersenne primes. The list continues. In 1963, Gillies found that the above number, $2^{9941} - 1$, is a Mersenne prime. With my laptop it takes less than 1 second for Python to check that this is indeed a prime number. The largest Mersenne prime found so far is $2^{82,589,933} - 1$, having 24,862,048 digits, which was discovered in January 2018.

Within integer arithmetic, there are division and remainder operations that Python can handle with // and %. It is a fact that for two non-zero integers n and m one can write $n = mq + r$, where q and r are unique integers with $0 \leq r < q$. In Python we get n//m=q and n % m =r. We check this with $n = 13$ and $m = 4$ below. Note the use of the function print in the example below.

```
[8]: print(13, '=', 13 // 3, 'x', 3 ,'+', 13 % 3)
```

```
[8]: 13 = 4 x 3 + 1
```

We will explore the function `print` more at the end of this chapter.

Exercise 1.1 *Show that the number 142857 is cyclic, meaning if we multiply it by any of the numbers 1 to 6, the result will be a permutation of the digits of the original number 142857.*

Solution

Later in Chapter 3 we will write a program to find cyclic numbers. Checking a number is cyclic, however, is quite easy:

```
[9]: print(142857*2, 142857*3, 142857*4,142857*5, 142857*6)
```

```
[9]: 285714 428571 571428 714285 857142
```

1.2 Integers, Reals and Complex Numbers

Python distinguishes between different classes of numbers. The *integers*, denoted by $\mathbb{Z} = \{\cdots, -2, -1, 0, 1, 2 \cdots\}$ in mathematics, are those numbers with no decimal points, such as all the numbers we have worked with so far, whereas *real numbers* or *floats*, denoted by $\mathbb{R}$, are those with decimal points, which are used for approximation, such as 1.2, $\sqrt{2}$ or π.

Python can also handle complex numbers using the format a+bj, where a is the real part and b is the imaginary part. Here $1j$ is what in the literature is commonly denoted by i, so that $1j^2 = -1$. But note how Python presents the complex numbers. In the example below we also add comments into the code. This is done by using # and writing the comment after it.

```
[10]: (1j)**2 # 1j represents the imaginary number i
```

```
[10]: (-1+0j)
```

As an example, we calculate $(\frac{1}{1} + \frac{1}{2} + \frac{1}{3}) \times 3$ and $(3 - 6i)^2$.

```
[11]: (1/1 + 1/2 + 1/3) * 3
```

```
[11]: 5.5
```

```
[12]: (3 - 6j)**2
```

[12]: (-27-36j)

Since $i^2 = -1$, it is easy to see $(a + bi)(a - bi) = a^2 + b^2$. We check this with an example.

[13]: `(2 - 3j)*(2 + 3j)`

[13]: (13+0j)

One of the reasons why Python has become so popular is that it has a large number of `libraries` which contain ready to use tools and functions. Various mathematical functions such as sin, cos or $\sqrt{}$ and log are available in different libraries of Python. The first library we will use is `math`, which makes many of these functions available to us. Note how we import this library into Python and how the functions (or methods) are used from this library. Once we have imported the library, we can access its built-in functions via dot ".".

[14]: `import math`

   `math.pi`

[14]: 3.141592653589793

[15]: `math.sin(math.pi/2)`

[15]: 1.0

Exercise 1.2 *Show that for any chosen angle x, Python gives* $\sin^2(x) + \cos^2(x) = 1$.

Solution

As an instance, choosing $\pi/5$ for the angle and translating the expression correctly into Python, we have

[16]: `math.sin(math.pi/5)**2 + math.cos(math.pi/5)**2`

[16]: 1.0

Of course, one could choose any other angle, run the code, and obtain 1 again.

Exercise 1.3 *Calculate the expression*

$$\sqrt[3]{e^\pi + \log\left(\frac{23}{\sin\left(\frac{\pi}{6}\right)}\right)}.$$

Solution

The only challenge here is to translate the mathematical expression correctly into Python.

```
[17]:  (math.exp(math.pi) + math.log(23 / math.sin(math.pi/6)))**(1/
    ↪3)
```

[17]: 2.998863793038475

Notice here that `math.exp(math.pi)` gives e^π.

Exercise 1.4 *Compute*

$$6 + \cfrac{1}{5 + \cfrac{1}{4 + \cfrac{1}{3 + \frac{1}{2}}}}.$$

Solution

One can immediately see a repeating pattern within this expression, and we will later write elegant codes to capture and compute such expressions. For the moment we can write the following. Note that the notation _ refers to the previous output in Jupyter. The symbol _ has other uses, as we will see throughout the book.

```
[18]:  3 + 1/2
```

[18]: 3.5

```
[19]:  4 + 1/_
```

[19]: 4.285714285714286

```
[20]:  5 + 1/_
```

[20]: 5.233333333333333

```
[21]:  6 + 1/_
```

[21]: 6.191082802547771

Or in one line as:

```
[22]:  6 + 1/(5 + 1/(4 + 1/(3 + 1/2)))
```

[22]: 6.191082802547771

Besides the `math` library, which provides many valuable mathematical functions, there are two other libraries in Python that are heavily used: numpy and sympy. The library numpy is designed for numerical computations whereas sympy is for symbolic computations and calculus. We look at these libraries in detail in Chapters 6 and 7. Here we just give an indication of how these libraries behave. All these libraries provide the basic mathematical functions, such as trigonometric functions sin, cos, tan, etc., however their methods of computation differ.

```
[23]:  import math
       import numpy
       import sympy
```

We will calculate $\sin(\pi/5)$ using each of these libraries.

```
[24]:  math.sin(math.pi/5)
```

[24]: 0.5877852522924731

```
[25]:  numpy.sin(numpy.pi/5)
```

[25]: 0.5877852522924731

```
[26]:  sympy.sin(sympy.pi/5)
```

[26]:
$$\sqrt{\frac{5}{8} - \frac{\sqrt{5}}{8}}$$

The sympy library gives $\sqrt{\frac{5}{8} - \frac{\sqrt{5}}{8}}$ as the value of $\sin(\pi/5)$, which is the exact value. This shows that sympy is not approaching the expressions numerically.

Next we will check the identity

$$\sin^2(\pi/5) + \cos^2(\pi/5) = 1$$

with the sympy functions.

```
[27]:  sympy.sin(sympy.pi/5)**2 + sympy.cos(sympy.pi/5)**2
```

[27]:
$$-\frac{\sqrt{5}}{8} + \frac{5}{8} + \left(\frac{1}{4} + \frac{\sqrt{5}}{4}\right)^2$$

We were expecting the output 1. In order to further simplify the output, we can use the function simplify in the sympy library.

```
[28]:  sympy.simplify(_)
```

[28]: 1

We will check the same identity within the numpy and math libraries.

```
[29]: numpy.sin(numpy.pi/5)**2 + numpy.cos(numpy.pi/5)**2
```

```
[29]: 1.0
```

```
[30]: math.sin(math.pi/5)**2 + math.cos(math.pi/5)**2
```

```
[30]: 1.0
```

We will dive into symbolic computations and sympy later on in this book. Here we just give a snippet of how to work with symbols. Using the library sympy we can introduce x as a symbol called **x**. Python can then carry out arithmetic symbolically with x, without enquiring what the value of x is.

```
[31]: x = sympy.symbols('x')
```

```
[32]: x + 1
```

```
[32]: x + 1
```

```
[33]: (2 * x + 3)**2
```

```
[33]: (2x + 3)^2
```

```
[34]: x = sympy.symbols('SometimesUPandsometimesDOWN')
```

```
[35]: x
```

$[35]:$ $SometimesUPandsometimesDOWN$

```
[36]: x /(1 + x)
```

$[36]:$ $$\frac{SometimesUPandsometimesDOWN}{SometimesUPandsometimesDOWN + 1}$$

We calculate the expression $\sin^2(x) + \cos^2(x)$ within sympy. Python then returns the correct identity of 1 for this expression.

```
[37]: sympy.simplify(sympy.sin(x)**2 + sympy.cos(x)**2)
```

```
[37]: 1
```

Exercise 1.5 *Investigate the following identities:*

$$\frac{(1 + \sqrt{5})^{10} - (1 - \sqrt{5})^{10}}{1024 \sqrt{5}} = 55.$$

Solution

We first use the library `math` to calculate the first expression.

```
[38]: ((1 + math.sqrt(5))**10 - (1 - math.sqrt(5))**10)/(1024*math.
      ↪sqrt(5))
```

```
[38]: 55.000000000000014
```

Note that we have got a floating point number which is *almost* 55. This is to be expected as the functions in the `math` library approach calculations "numerically" and these methods will not give exact results. However the results are generally extremely good with high precision.

Approaching the computations with numpy built-in functions would give the same approximation.

```
[39]: ((1 + numpy.sqrt(5))**10 - (1-numpy.sqrt(5))**10)/
      ↪(1024*numpy.sqrt(5))
```

```
[39]: 55.000000000000014
```

Employing `sympy` capabilities, we can actually show that equality holds for this identity.

```
[40]: ((1 + sympy.sqrt(5))**10 - (1 - sympy.sqrt(5))**10)/
      ↪(1024*sympy.sqrt(5))
```

$$[40]: \quad \frac{\sqrt{5}\left(-\left(1 - \sqrt{5}\right)^{10} + \left(1 + \sqrt{5}\right)^{10}\right)}{5120}$$

```
[41]: sympy.simplify(_)
```

```
[41]: 55
```

We emphasise that floats (real numbers) are all about approximations.

```
[42]: 0.1 + 0.2
```

```
[42]: 0.30000000000000004
```

```
[43]: 0.3 - (0.1 + 0.2)
```

```
[43]: -5.551115123125783e-17
```

Although we were expecting 0, we got a non-zero number which is almost zero. Using the built-in function `round` we can comfortably see what we got is indeed extremely close to zero.

```
[44]: round(0.3 - (0.1 + 0.2))
```

```
[44]: 0
```

Because Python is so widely used, help, hints and good samples are always a google away. But one can also get a summary of the functions by using the command help.

```
[45]: help(round)
```

```
[45]: Help on built-in function round in module builtins:

      round(number, ndigits=None)
          Round a number to a given precision in decimal digits.

          The return value is an integer if ndigits is omitted or
          None.  Otherwise the return value has the same type as the
          number.  ndigits may be negative.
```

1.3 Objects and their Types

Everything in Python is an object. One can think of an object as an ecosystem with its own data and its own tools and functions that can be used to modify the data. Even the numbers 3, 3.5 and $3.2 - 2i$ in Python are all objects. The objects have different types. The function type determines the type of the objects we are working with:

```
[46]: type(3)
```

```
[46]: int
```

```
[47]: type(math.sqrt(2))
```

```
[47]: float
```

```
[48]: type(3.2 - 2j)
```

```
[48]: complex
```

Once we have an object, we can access their methods and functions via dot ".". As an example, the float object has a method called is_integer, which determines if a real number is indeed an integer.

```
[49]:  x = math.sqrt(16)
```

```
[50]:  x.is_integer()
```

```
[50]:  True
```

Here the parameter x is assigned to math.sqrt(16). So x is an object of type float. We can then use the methods and functions that come with this object, one being is_integer.

Exercise 1.6 *Let m be a natural number and*

$$A = \frac{(m+3)^3 + 1}{3m}.$$

Find all the integers m less than 10 such that A is an integer. Show that A is always odd.

Solution

Although we don't currently have many tools at our disposal, we can translate the formula for A in Python, replace m by 1, 2, ... and each time check if the computation returns an integer with the method is_integer(). Later on, when we know how to create loops, we will revisit this exercise in Chapter 5 (Exercise 5.7) and find all the numbers up to 500.

```
[51]:  m = 2
       x = ((m + 3)**3 + 1)/(3 * m)
       x.is_integer()
```

```
[51]:  True
```

```
[52]:  x
```

```
[52]:  21.0
```

When working with a complex number $a + bi$, we would always like to have access to the *real* part, a, and the *imaginary* part, b. The complex object comes with methods to obtain this information.

```
[53]:  x = 3.4 - 5.0j
```

```
[54]:  print(x.real, '  ', x.imag)
```

```
[54]:  3.4     -5.0
```

With `x.imag` we can access the imaginary part of the complex number x. This imaginary part is a float object which comes with its own methods, which we now use.

```
[55]: x.imag.is_integer()
```

```
[55]: True
```

```
[56]: x.real.is_integer()
```

```
[56]: False
```

Although the library `math` provides basic functions, such as `sin` and `cos`, these functions are designed to handle real (float) numbers. If we are working with complex numbers, we need to import the library `cmath` and its functions, which allow us to handle complex arithmetic.

Exercise 1.7 *If a and b are real numbers, show that the real part of*

$$\left(\cos(a + bi) + \sin(a + bi) \right)^2$$

is equal to

$$1 + \sin(2a)\left(\cosh(b)^2 + \sinh(b)^2 \right)$$

and find a similar expression for its imaginary part.

Solution

Although this identity should be valid for *any* a and b, we are going to choose some values for a and b and test the claim. Later when we define functions in Python we shall be able to check this for any value of a and b. Since we are working with complex numbers, we import the library `cmath`.

```
[57]: import cmath

      (cmath.cos(1 + 1j) + cmath.sin(1 + 1j))**2
```

```
[57]: (4.420954861117013-1.5093064853236153j)
```

```
[58]: ((cmath.cos(1 + 1j) + cmath.sin(1 + 1j))**2).real
```

```
[58]: 4.420954861117013
```

```
[59]: ((cmath.cos(1 + 1j) + cmath.sin(1 + 1j))**2).imag
```

```
[59]: -1.5093064853236153
```

```
[60]:  1 + cmath.sin(2)*(cmath.cosh(1)**2 + cmath.sinh(1)**2)
```

[60]: (4.420954861117013+0j)

Comparing the results we see the statement of the exercise is valid for this particular *a* and *b*.

1.4 Importing Libraries in Python

We have seen that `import math` makes the library `math` available for use. If the name of a library is long, one can introduce an alias.

```
[61]:  import numpy as np
```

```
[62]:  np.sin(np.pi / 2)
```

[62]: 1.0

Although one could use `import numpy as MyDarling` or any other alias, the name `np` has become quite widespread in the community.

If one uses certain methods of a library quite often, one can import them directly.

```
[63]:  from sympy import sin, cos, pi, simplify, symbols
```

```
[64]:  x = symbols('x')
        y = symbols('what?')
```

```
[65]:  (x + 1/y) * (1/x + y)
```

$$[65]:\ \left(what? + \frac{1}{x}\right)\left(x + \frac{1}{what?}\right)$$

```
[66]:  simplify(sin(x)**2 + cos(x)**2)
```

[66]: 1

We should warn the reader that confusion might arise as to where these methods belong. If one imports methods from different libraries then the later methods will overwrite the earlier ones.

1.5 Variables in Python

In order to feed data into a computer program one needs to define variables to be able to assign data to them. Python's methodology in assigning data to variables is slightly different than in other languages. We start with the simplest assignment of a number to a variable.

```
[67]: x = 2
```

```
[68]: x
```

```
[68]: 2
```

As long as you use common sense, any names you choose for variables are valid (provided they are not used as Python keywords). Names like x, y, x3, myfunc, xQuaternion as well as x_3, my_func, x_Quaternion are all fine. Also note that Python is case-sensitive, thus xy and xY are considered as two different variables. We assign values to these variable using ; to write the code in one line.

```
[69]: xy = 12 ; xY = -1.4; x_y = 2-3j
```

```
[70]: xy
```

```
[70]: 12
```

```
[71]: xY
```

```
[71]: -1.4
```

We could also assign the values in line as a tuples. We will see these concepts in Chapter 2.

```
[72]: xy, xY, x_y = 12, -1.4, 2-3j
```

```
[73]: xY
```

```
[73]: -1.4
```

```
[74]: xy**2 - xY**2 + x_y**2
```

```
[74]: (137.04-12j)
```

One crucial difference in Python that we need to understand at this early stage is that the variables are *pointers*. In other languages the command x = 2 creates a cell or (an object) labelled x and 2 is stored in that cell, whereas in Python there is an integer object 2 and x is a pointer, pointing to that object. There is no immediate harm if we just think of x as a variable that 2 is assigned to, but in the background we have to be aware of the Python philosophy.

```
[75]: x = y = z = 10
```

The way we should understand this line is that we have created three pointers all pointing to the integer object 10. We could change one of the pointers, pointing to a different object without changing the direction of the other two pointers.

```
[76]: z = 12
```

We print the values of these variables. Note that the command \n in print breaks the line as the result shows.

```
[77]: print('x is pointing to', x, '\ny is pointing to', y, '\nz is
      ⇔pointing to', z)
```

```
[77]: x is pointing to 10
      y is pointing to 10
      z is pointing to 12
```

This becomes more clear when we work with objects that can be changed (mutable), for example *lists*.

On some occasions the existence of a variable is needed more so than the variable itself. Sometimes we require that a *dummy variable* runs through a loop, or a list. In this case we can simply use _ instead of coming up with a name.

```
[78]: _ = 5
```

```
[79]: _
```

```
[79]: 5
```

```
[80]: a, _ , _, b = 10, 13, 12, 100
```

```
[81]: print(a,'and', b)
```

```
[81]: 10 and 100
```

1.6 Equalities and Boolean Statements

Primarily there are two equalities in Python = and ==. The first one creates a pointer, whereas == is used for comparison. The result of the comparison is a boolean value of True or False

```
[82]:  1 + 2 == 3
```

```
[82]:  True
```

```
[83]:  3**2 + 4**2 == 5**2
```

```
[83]:  True
```

```
[84]:  9**10 == 10**9
```

```
[84]:  False
```

Recall the Euclidean division $n = mq + r$. We can check its validity here with an example.

```
[85]:  23 == (23 // 4) * 4  + 23 % 4
```

```
[85]:  True
```

However we should be careful when working with floats, as the computations have been done up to a certain precision.

```
[86]:  0.1 + 0.2 == 0.3
```

```
[86]:  False
```

```
[87]:  0.1 + 0.2
```

```
[87]:  0.30000000000000004
```

In mathematical logic, statements can have a value of True, False or undefined. These are called *Boolean expressions*. This helps us to "make a decision" and write programs based on the value of a statement. We will see later how to use if-else statements to control the flow of the program based on the value of boolean expressions.

One can combine logical statements with the usual boolean operations and, or, not or the equivalent &, |, !, as the following examples show:

```
[88]:  2 > 3
```

```
[88]:  False
```

```
[89]:  not(2 > 3)
```

```
[89]:  True
```

[90]: ```
2 > 3 or 3 > 2
```

[90]: True

[91]: ```
3**2 + 4**2 >= 5**2
```

[91]: True

Exercise 1.8 *Is the following statement correct!?*

```
(1 < 2 < 3) == (1 < 2) and (2 < 3)
```

True

Solution

In fact the way this statement is written is misleading! Indeed $1 < 2 < 3$ means exactly $1 < 2$ and $2 < 3$. However in the statement above, Python evaluates the left-hand side (which is True) and then evaluates the right-hand side (again True) and then compare the two. Therefore the statement below also gives True, which is not mathematically equivalent.

[92]: ```
(1 < 2 < 3) == (1 < 2) and (2 < 10)
```

[92]: True

**Exercise 1.9** *The hyperbolic functions are combinations of exponential functions. As an example*

$$\sinh(x) = \frac{e^x - e^{-x}}{2}, \text{ and } \cosh(x) = \frac{e^x + e^{-x}}{2}$$

*using the* math *library check that these equalities hold for any x.*

*Solution*

The hyperbolic function sinh is available in the math library. We will check if this function returns the same value as $\frac{e^x - e^{-x}}{2}$, here for $\pi$. Later in Chapter 4, we define functions within Python. Once this is done, we can define the function $f(x) = \frac{e^x - e^{-x}}{2}$ and systematically compare these two functions.

[93]: ```
import math

p = math.pi
sh = (math.exp(p) - math.exp(-p))/2
math.sinh(p) == sh
```

```
[93]: False
```

```
[94]: print(math.sinh(p), sh)
```

```
[94]: 11.548739357257746 11.548739357257748
```

Again, we can see that, although the results are not precisely the same, they are very very close!

1.7 Strings

We have seen several types of objects: integers, floats, complex as well as booleans. We finish this chapter with the type `string`.

```
[95]: 'this is a string'
```

```
[95]: 'this is a string'
```

```
[96]: message = 'Western Sydney '
      type(message)
```

```
[96]: str
```

One can perform certain 'arithmetic' operations with strings, as the following examples show.

```
[97]: message + 'University'
```

```
[97]: 'Western Sydney University'
```

```
[98]: message * 3
```

```
[98]: 'Western Sydney Western Sydney Western Sydney '
```

```
[99]: message
```

```
[99]: 'Western Sydney '
```

```
[100]: message = message + 'University'
```

```
[101]: message
```

```
[101]: 'Western Sydney University'
```

1.8 Strings as Objects

We have already mentioned that everything in Python is an object. Once the object is defined, the functions and methods are all at our disposal to use. Here we will use several methods available for str objects. The examples below show how these methods are used.

```
[102]: message
```

```
[102]: 'Western Sydney University'
```

```
[103]: message.capitalize()
```

```
[103]: 'Western sydney university'
```

```
[104]: message.lower()
```

```
[104]: 'western sydney university'
```

```
[105]: message.upper()
```

```
[105]: 'WESTERN SYDNEY UNIVERSITY'
```

```
[106]: message
```

```
[106]: 'Western Sydney University'
```

All the methods used above, upper, lower, capitalize are provided within the object of string. As is clear from the example, they operate on the object message but do not change the object itself.

One very useful method when working with strings is to use *format strings*, which allows us to pass data inside a string where the location is determined by {}.

```
[107]: 'University of {}'.format('Sydney')
```

```
[107]: 'University of Sydney'
```

```
[108]: 'University of {}'.format(123)
```

```
[108]: 'University of 123'
```

```
[109]: name = 'First name: {}, Last name: {}'
```

```
[110]: name.format('Lustig', 'Sabzian')
```

[110]: `'First name: Lustig, Last name: Sabzian'`

There is a simpler way to use *format strings*, as the following shows. They can be used very nicely with the command print, as we will see later on.

[111]:
```
fn = 'Lustig'
ln = 'Sabzian'

f'First name: {fn}, last name: {ln}'
```

[111]: `'First name: Lustig, last name: Sabzian'`

[112]:
```
x = 2
y = math.sqrt(x)

f'the square root of {x} is {y}'
```

[112]: `'the square root of 2 is 1.4142135623730951'`

1.9 Input and Output

We close this chapter with two useful commands. The command input asks for data to be given to the program by the user. The data entered is captured as a string.

[113]:
```
s = input('enter a text: ')
print(s + ' checked')
```

enter a text: Python

[113]: Python checked

Here is a standard way to change the data to numbers, in case one requires the input to be integers, or reals etc.

[114]:
```
s = int(input('enter a number: '))
s
```

enter a number: 666

[114]: 666

[115]:
```
s = input('Name ')
t = input('age ')
s + ' is ' + t + ' years old.'
```

```
Name Dad
age 92
```

[115]: 'Dad is 92 years old.'

The command print will create an output. The following examples show the various ways one can use print.

[116]: `print('printing a string')`

[116]: printing a string

[117]: `print(2 * 3 * 4 + 1.4)`

[117]: 25.4

[118]: `print(12, 'is smaller than', 15)`

[118]: 12 is smaller than 15

As mentioned, the format strings fit very nicely with the print command.

[119]:
```
x = 2
y = math.sqrt(x)
print(f'the square root of {x} is {y}')
```

[119]: the square root of 2 is 1.4142135623730951

Finally there are certain special characters that can be used with strings. Among them are \n and \t, which give a new line and a new tab, respectively.

[120]: `print('Hello\tworld\nfinal\tgreetings')`

[120]:
```
Hello    world
final    greetings
```

We finish the chapter with an amusing example, showcasing several of the methods available for working with strings.

[121]:
```
secret = 'xzwoy thx uilxzcx'
x = 'zxawu'
y = 'neojs'
table = secret.maketrans(x, y)
print(secret.translate(table))
```

[121]: enjoy the silence

Problems

1) Compute

$$\sqrt{1 + \frac{1}{1^2} + \frac{1}{2^2}} + \sqrt{1 + \frac{1}{2^2} + \frac{1}{3^2}} + \sqrt{1 + \frac{1}{3^2} + \frac{1}{4^2}}$$

2) Compute

$$\left(\frac{1}{2} + \frac{1}{3} + \frac{1}{5} + \frac{1}{7}\right) + \left(\frac{5}{2} + \frac{5}{3} + \frac{5}{5} + \frac{5}{7}\right) + \left(\frac{11}{2} + \frac{11}{3} + \frac{11}{5} + \frac{11}{7}\right)$$

3) Show that for any two positive numbers a and b, if $a + b = 1$, then

$$\left(a + \frac{1}{a}\right)^2 + \left(b + \frac{1}{b}\right)^2 \geq \frac{25}{2}.$$

4) Use Python to show that

$$\tan\frac{3\pi}{11} + 4\sin\frac{2\pi}{11} = \sqrt{11}.$$

Note, one needs to use the math library or numpy to have access to the square root, sine and tangent functions.

5) Show that

$$\sqrt{\sqrt[3]{64}(2^2 + (1/2)^2)} - 1 = 4.$$

6) Show that

$$\sin(\frac{2\pi}{10}) + \sin(\frac{4\pi}{10}) + \cdots + \sin(\frac{10\pi}{10}) = \sqrt{\frac{1}{2}\left(5 - \sqrt{5}\right)} + \sqrt{\frac{1}{2}\left(5 + \sqrt{5}\right)}$$

7) Show that

$$\left(\frac{1}{2} + \cos\left(\frac{\pi}{20}\right)\right)\left(\frac{1}{2} + \cos\left(\frac{3\pi}{20}\right)\right)\left(\frac{1}{2} + \cos\left(\frac{9\pi}{20}\right)\right)\left(\frac{1}{2} + \cos\left(\frac{27\pi}{20}\right)\right) = \frac{1}{16}.$$

8) By looking at various examples, observe that the product of four consecutive numbers plus one is always a square number.

9) Using Python, demonstrate that

$$\frac{1 + \sin(x) - \cos(x)}{1 + \sin(x) + \cos(x)} = \tan(x/2).$$

Chapter 2
Lists and Tuples

2.1 Data: Lists

In Python, lists provide the first basic building blocks for working with and handling data.

One can think of a computer program as a function which accepts some (crude) data or information and gives back the data we would like to obtain. The classic Python provides certain tools to collect and handle data, such as `list`, `tuple`, `set` and `dictionary`. Libraries such as `numpy` further provide capable tools to work with a large collection of data. In this chapter we will study `list` and `tuple`; how to collect data and how to access and process the elements in these collections. We will study `set` and `dictionary` in Chapter 5.

Data Science starts with handling data, such as data cleansing, aggregation, transformation, and data visualisation, and lists are the first stop in this process. Once we grasp the concept of the lists and how to work with them, we can comfortably work with other objects which are designed for handling data.

We start with an example of a `list`:

```
[1]: L = [3, 6.4, 3, 'stuff', 'x^2+x+1', 64/3]
```

```
[2]: type(L)
```

```
[2]: list
```

```
[3]: L
```

```
[3]: [3, 6.4, 3, 'stuff', 'x^2+x+1', 21.333333333333332]
```

© The Author(s), under exclusive license to Springer Nature Switzerland AG 2023
R. Hazrat, *A Course in Python*, Springer Undergraduate Mathematics Series,
https://doi.org/10.1007/978-3-031-49780-3_2

list is one of the ways to collect data in Python. As this example shows, the data (of any type and format) are arranged between square brackets, [and]. Lists respect order and repetition of the data:

```
[4]: [1, 2] == [2, 1]
```

```
[4]: False
```

```
[5]: [1, 1, 2] == [1, 2]
```

```
[5]: False
```

We can check if an element belongs to a list

```
[6]: 2 in [1, 2, 3]
```

```
[6]: True
```

```
[7]: 'mode' in ['fast', 'fashion']
```

```
[7]: False
```

Similar to strings, we can do certain arithmetic operations with lists. Recall the list L above.

```
[8]: L + ['more', 'less', [4, 4, 4]]
```

```
[8]: [3, 6.4, 3, 'stuff', 'x^2+x+1', 21.333333333333332, 'more',
      'less', [4, 4, 4]]
```

```
[9]: ['Western', 'Sydney'] * 2
```

```
[9]: ['Western', 'Sydney', 'Western', 'Sydney']
```

```
[10]: [1, 2] + [3, 4]
```

```
[10]: [1, 2, 3, 4]
```

As the above examples show, we can add another list to the end of a given list by adding them together or repeat the list by multiplication.

2.1.1 Accessing entries of a list

It is natural to need to access the elements of a list.

```
[11]: L
```

```
[11]: [3, 6.4, 3, 'stuff', 'x^2+x+1', 21.333333333333332]
```

```
[12]: L[0]
```

```
[12]: 3
```

```
[13]: L[4]
```

```
[13]: 'x^2+x+1'
```

```
[14]: L[-1]
```

```
[14]: 21.333333333333332
```

```
[15]: L[-2]
```

```
[15]: 'x^2+x+1'
```

In Python indexing of elements starts from 0. Thus we start counting from 0 (which refers to the first element of the list!) Examining the above examples reveals that L[i] gives the $i + 1$-th member of the list. Thus L[0] points to the first element in the list. To obtain consecutive elements of a list one can use the command L[n : m]. Here L[n : m] retrieves the elements in the list L, from the $n + 1$-th up to (at most) the m-th item. The command L[n : m : s] introduces "step" s, and one retrieves elements $n + 1, n + 1 + s, n + 1 + 2s, \ldots$ up to the m-th element. The examples below make this clear.

```
[16]: L
```

```
[16]: [3, 6.4, 3, 'stuff', 'x^2+x+1', 21.333333333333332]
```

```
[17]: L[1 : 5]
```

```
[17]: [6.4, 3, 'stuff', 'x^2+x+1']
```

Here is the meaning of retrieving up to *at most* the m-th item in L[n:m].

```
[18]: L[1 : 100]
```

```
[18]: [6.4, 3, 'stuff', 'x^2+x+1', 21.333333333333332]
```

```
[19]: L[1 : 5 : 2]
```

```
[19]: [6.4, 'stuff']
```

Notice that in the above examples, we start with the second element in the list and go all the way *up to* the sixth element and each time pick every second element.

In the command L[n : m], if we leave out n, the list starts from the beginning, and if we leave out m, it goes all the way to the end. The examples below make this clear.

```
[20]: L
```

[20]: [3, 6.4, 3, 'stuff', 'x^2+x+1', 21.333333333333332]

```
[21]: L[ : 5]
```

[21]: [3, 6.4, 3, 'stuff', 'x^2+x+1']

```
[22]: L[ : -1]
```

[22]: [3, 6.4, 3, 'stuff', 'x^2+x+1']

```
[23]: L[4 : ]
```

[23]: ['x^2+x+1', 21.333333333333332]

```
[24]: L[ : 100]
```

[24]: [3, 6.4, 3, 'stuff', 'x^2+x+1', 21.333333333333332]

```
[25]: L[4 : 1 : -1]
```

[25]: ['x^2+x+1', 'stuff', 3]

```
[26]: L[ : : -1]
```

[26]: [21.333333333333332, 'x^2+x+1', 'stuff', 3, 6.4, 3]

```
[27]: L == L[ : 4] + L[4 : ]
```

[27]: True

Exercise 2.1 *Define the lists*

```
campus=['Parramatta', 'Campbelltown', 'Kingswood']
```

and allocation=[30,10,7]

and then create the list

```
total=['Parramatta', 30, 'Campbelltown', 10, 'Kingswood, 7]
```

Solution

We first define our given lists.

```
[28]:  campus = ['Parramatta ', ' Campbelltown ', ' Kingswood ']
```

```
[29]:  allocation = [30, 10, 7]
```

This is the most naive way to put these lists together:

```
[30]:  total = [[campus[0] , allocation[0]] ,
                [campus[1] , allocation[1]] , [campus[2] ,
          ⌐allocation[2]]]
```

```
[31]:  total
```

```
[31]:  [['Parramatta ', 30], [' Campbelltown ', 10],
          [' Kingswood ', 7]]
```

We could also use the following rather more clever approach:

```
[32]:  CA = campus + allocation
```

```
[33]:  CA
```

```
[33]:  ['Parramatta ', ' Campbelltown ', ' Kingswood ', 30, 10, 7]
```

```
[34]:  [CA[0 :: 3], CA[1 :: 3], CA[2 :: 3]]
```

```
[34]:  [['Parramatta ', 30], [' Campbelltown ', 10],
          [' Kingswood ', 7]]
```

Yet another way to put these together (in the absence of loops):

```
[35]:  CA = []
```

```
[36]:  CA += [[campus[0], allocation[0]]]
       CA += [[campus[1], allocation[1]]]
       CA += [[campus[2], allocation[2]]]

       CA
```

```
[36]:  [['Parramatta ', 30], [' Campbelltown ', 10],
          [' Kingswood ', 7]]
```

Of course, for a larger collection of lists, and with the loop facilities at our disposal (which we will meet in Chapter 3), the above code could be modified to pair the lists together via a loop.

Returning to our list L, we can replace an element inside the list by singling out that element and assigning new data to it.

```
[37]: L
```

```
[37]: [3, 6.4, 3, 'stuff', 'x^2+x+1', 21.333333333333332]
```

```
[38]: L[4] = 'replacement'
```

```
[39]: L
```

```
[39]: [3, 6.4, 3, 'stuff', 'replacement', 21.333333333333332]
```

```
[40]: L[0] = L[2] = L[4] = 'XXX'
```

```
[41]: L
```

```
[41]: ['XXX', 6.4, 'XXX', 'stuff', 'XXX', 21.333333333333332]
```

One of the secrets of writing code comfortably is that one should be able to manipulate lists easily. Often in applications, situations like the following arise:

- Given $\{x_1, x_2, \cdots, x_n\}$ and $\{y_1, y_2, \cdots, y_n\}$, produce

$$\{x_1, y_1, x_2, y_2, \cdots, x_n, y_n\},$$

and

$$\{\{x_1, y_1\}, \{x_2, y_2\}, \cdots, \{x_n, y_n\}\}.$$

- Given $\{x_1, x_2, \cdots, x_n\}$ and $\{y_1, y_2, \cdots, y_n\}$, produce

$$\{x_1 + y_1, x_2 + y_2, \cdots, x_n + y_n\}.$$

- Given $\{x_1, x_2, \cdots, x_n\}$ and $\{y_1, y_2, \cdots, y_n\}$, produce

$$\{\{x_1, y_1\}, \{x_1, y_2\}, \cdots, \{x_1, y_n\}, \{x_2, y_1\}, \{x_2, y_2\}, \cdots, \{x_2, y_n\},$$
$$\cdots, \{x_n, y_1\}, \{x_n, y_2\}, \cdots, \{x_n, y_n\}\}.$$

- Given $\{x_1, x_2, \cdots, x_n\}$, produce

$$\{x_1, x_1 + x_2, \cdots, x_1 + x_2 + \cdots + x_n\}.$$

- Given $\{x_1, x_2, \cdots, x_n\}$, produce

$$\Big\{\{\{x_1\}, \{x_2, \ldots, x_n\}\}, \{\{x_1, x_2\}, \{x_3, \ldots, x_n\}\} \ldots \{\{x_1, \ldots x_{n-1}\}, \{x_n\}\}\Big\}.$$

As we progress, we will see Python and especially the library numpy provide tools to produce such combinations from lists and other collections of data.

2.2 Lists as Objects

Recall that everything in Python is an object, which comes with its own tools (i.e., methods). With this view, lists are objects which we can change, the so-called *mutable* objects. The objects that cannot be changed are called *immutable*. We have already seen one object that, once created, cannot be modified, namely the object integers.

We will demonstrate how to use some of the list methods to modify a given list.

```
[42]:  L = [3, 6.4, 3, 'stuff', 'x^2+x+1', 64/3]
```

```
[43]:  L.append('extra bit')
```

```
[44]:  L
```

```
[44]:  [3, 6.4, 3, 'stuff', 'x^2+x+1', 21.333333333333332,
          'extra bit']
```

append is one of the methods available in the object list. It adds an item to the end of the list, as the example above demonstrates. Already the method append allows us to solve the above exercise differently.

```
[45]:  total = [];
       total.append(campus[0])
       total.append(allocation[0])
```

```
[46]:  total
```

```
[46]:  ['Parramatta ', 30]
```

```
[47]:  total.append(campus[1]); total.append(allocation[1])
```

```
[48]:  total
```

```
[48]:  ['Parramatta ', 30, ' Campbelltown ', 10]
```

[49]: ```
total.append(campus[2]); total.append(allocation[2])
```

[50]: ```
total
```

[50]: `['Parramatta ', 30, ' Campbelltown ', 10, ' Kingswood ', 7]`

Besides append, list comes with several other methods, such as count, reverse, ... As an example:

[51]: ```
L = [1, 2, 1.2, 3 - 1, 'stuff', 2, 4]
L.count(2)
```

[51]: 3

[52]: ```
L.reverse()
L
```

[52]: `[4, 2, 'stuff', 2, 1.2, 2, 1]`

Lists can have other lists as elements. We can envisage the notion of matrices in mathematics via lists of list. Recall that an $n \times m$ matrix is mathematical objects with n-rows and m-columns. An $n \times m$ matrix A then consists of $n \times m$ objects (usually numbers) and an entry on the i-th row and j-column can be denoted by a_{ij}. We translate the 3×3 matrix A below into Python

$$A = \begin{pmatrix} 1 & 2 & 3 \\ 4 & 5 & 6 \\ 7 & 8 & 9 \end{pmatrix}.$$

The example below can be thought of as a matrix with three rows and three columns

[53]: ```
L = [[1, 2, 3], [4, 5, 6], [7, 8, 9]]
```

[54]: ```
L
```

[54]: `[[1, 2, 3], [4, 5, 6], [7, 8, 9]]`

[55]: ```
L[0]
```

[55]: `[1, 2, 3]`

[56]: ```
L[0][0]
```

[56]: 1

```
[57]:  L[0][0] = L[1][1] = L[2][2] = 'Diagonal'
```

```
[58]:  L
```

```
[58]:  [['Diagonal', 2, 3], [4, 'Diagonal', 6], [7, 8, 'Diagonal']]
```

Exercise 2.2 *Create a 4 × 3 matrix with entries all 0*

$$A = \begin{pmatrix} 0\,0\,0 \\ 0\,0\,0 \\ 0\,0\,0 \\ 0\,0\,0 \end{pmatrix}.$$

Solution

Here is a clever way to do this, benefiting from the arithmetic on lists.

```
[59]:  S = [[0] * 3] * 4
       S
```

```
[59]:  [[0, 0, 0], [0, 0, 0], [0, 0, 0], [0, 0, 0]]
```

One needs to be a bit careful here: when we use list * 3, Python creates three copies of the same object. This becomes clear when we change one of the objects, as the example below shows:

```
[60]:  S[0][1] = 'upset'
       S
```

```
[60]:  [[0, 'upset', 0], [0, 'upset', 0], [0, 'upset', 0],
        [0, 'upset', 0]]
```

Compare the above with the following:

```
[61]:  T = [[0, 0, 0], [0, 0, 0], [0, 0, 0], [0, 0, 0]]
```

```
[62]:  T[0][1] = 'upset'
       T
```

```
[62]:  [[0, 'upset', 0], [0, 0, 0], [0, 0, 0], [0, 0, 0]]
```

```
[63]:  S = [[0] * 3] * 3
       S
```

```
[63]:  [[0, 0, 0], [0, 0, 0], [0, 0, 0]]
```

```
[64]: S[1] = [-1, -2]
      S
```

```
[64]: [[0, 0, 0], [-1, -2], [0, 0, 0]]
```

Of course one needs to be mindful of what one adds to a list and how to handle it. We give a curious example and leave it to the reader to further explore this direction.

```
[65]: L=[3, 6.4, 3, 'stuff', 'x^2+x+1', 64/3]
```

```
[66]: L.append(L)
```

```
[67]: print(L)
```

```
      [3, 6.4, 3, 'stuff', 'x^2+x+1', 21.333333333333332, [...]]
```

```
[68]: L[-1]
```

```
[68]: [3, 6.4, 3, 'stuff', 'x^2+x+1', 21.333333333333332, [...]]
```

```
[69]: L[-1][0]
```

```
[69]: 3
```

```
[70]: L[-1][-1]
```

```
[70]: [3, 6.4, 3, 'stuff', 'x^2+x+1', 21.333333333333332, [...]]
```

```
[71]: L[-1][-1][-1]
```

```
[71]: [3, 6.4, 3, 'stuff', 'x^2+x+1', 21.333333333333332, [...]]
```

We can now explore the notion of pointers (variables) and mutable objects a bit more using lists. Suppose we define two pointers both pointing to a list.

```
[72]: x = y = [1, 2, 3, 4]
```

```
[73]: x
```

```
[73]: [1, 2, 3, 4]
```

```
[74]: y
```

```
[74]: [1, 2, 3, 4]
```

Now if we modify the object [1, 2, 3, 4] then both pointers x and y which are pointing to this object will show this change

```
[75]: x.append('extra bit')
      x
```

```
[75]: [1, 2, 3, 4, 'extra bit']
```

```
[76]: y
```

```
[76]: [1, 2, 3, 4, 'extra bit']
```

However we can re-direct the pointer y to a different object.

```
[77]: y = ['another', 'object']
      y
```

```
[77]: ['another', 'object']
```

```
[78]: x
```

```
[78]: [1, 2, 3, 4, 'extra bit']
```

In order to show how versatile lists are, we use the Python library PIL, which handles images. Recall we can import libraries into Python as follows:

```
[79]: from PIL import Image
```

```
[80]: x = Image.open('dog.jpg')
      y = Image.open('Napoleon.jpg')
```

The files dog.jpg and Napoleon.jpg are in the local directory. We have now assigned x and y to these pictures, respectively.

```
[81]: x
```

```
[81]:
```

[82]: `y`

[82]:

Now we can define a list which contains different types of objects, including images

[83]: `pic = ['cute dog', x , x.size, 'Napoleon', y, y.size]`

[84]: `display(pic[0], pic[1], pic[2])`

[84]: `'cute dog'`

`(144, 178)`

The object image comes with its own methods and tools. We just sample one or two of them here.

[85]: `pic[4].rotate(180)`

[85]:

[86]: `pic[1].effect_spread(10)`

[86]:

Exercise 2.3 *Given a list, swap the first and the last element of the list.*

Solution

First, we define a list to work with.

[87]: `L=['first', 4, 8, 'stuff', 9/2, 'last']`

Of course one crude approach is to do this:

[88]:
```
L[0] = 'last'
L[-1] = 'first'
```

```
[89]: L
```

```
[89]: ['last', 4, 8, 'stuff', 4.5, 'first']
```

or a slightly better approach:

```
[90]: temp = L[0]; L[0] = L[-1]; L[-1] = temp
```

```
[91]: L
```

```
[91]: ['first', 4, 8, 'stuff', 4.5, 'last']
```

However there is a more elegant way to do this:

```
[92]: L[0], L[-1] = L[-1], L[0]
```

```
[93]: L
```

```
[93]: ['last', 4, 8, 'stuff', 4.5, 'first']
```

and finally there is yet another elegant way using the sequences:

```
[94]: [f, *r , l] = L
```

```
[95]: [l, *r , f]
```

```
[95]: ['first', 4, 8, 'stuff', 4.5, 'last']
```

Here f is assigned to the first element of the list L, l is assigned to the last element and *r assigned to the sequence between the first and last element. Thus in the next line [l, *r, f] we simply swap the first and last element and keep the sequence *r. Python can even recognise the assignment if we drop the brackets:

```
[96]: f, *r , l = L
```

```
[97]: [l, *r, f]
```

```
[97]: ['first', 4, 8, 'stuff', 4.5, 'last']
```

Exercise 2.4 *Consider the matrix*

$$M = \begin{pmatrix} a & b & c \\ d & e & f \\ g & h & k \end{pmatrix}$$

and create its transpose.

Solution

We will later see that the library numpy has been designed to work seamlessly with matrices. Here using classical Python tools, we can write.

```
[98]: m = [['a', 'b', 'c'], ['d', 'e', 'f'], ['g', 'h', 'k']]
```

```
[99]: n = [[0, 0, 0], [0, 0, 0], [0, 0, 0]]
```

```
[100]: n[0][0], n[1][0], n[2][0] = m[0]
```

```
[101]: n[0][1], n[1][1], n[2][1] = m[1]
```

```
[102]: n[0][2], n[1][2], n[2][2] = m[2]
```

```
[103]: n
```

```
[103]: [['a', 'd', 'g'], ['b', 'e', 'h'], ['c', 'f', 'k']]
```

Before we explore other data structures in Python, we re-visit strings. One can access the elements within a string in a similar manner as lists

```
[104]: s = 'To be born again'
```

```
[105]: s[0]
```

```
[105]: 'T'
```

```
[106]: s[1]
```

```
[106]: 'o'
```

```
[107]: s[ : 5]
```

```
[107]: 'To be'
```

```
[108]: s[5 : ]
```

```
[108]: ' born again'
```

```
[109]: s[ : : -1]
```

```
[109]: 'niaga nrob eb oT'
```

```
[110]: s = 'kayak'
```

```
[111]: s[ : : -1] == s
```

```
[111]: True
```

2.3 A First Glimpse of Graphics

One of the most used libraries in Python is `matplotlib`. This library is used for plotting data and creating professional and magnificent two-dimensional graphics. We will devote an entire Chapter 7 to this library. Here we give the first instances of how `list` along with `matplotlib` allow us to plot data. First, we import the library into Python.

```
[112]: import matplotlib.pyplot as plt
```

In order to plot a graph with `matplotlib`, we specify the x-coordinates and then the y-coordinates. If `x=[x1,x2,..., xn]` and `y=[y1, y2,..., yn]`, then `plt.plot(x,y)` will produce a graph determined by pairs $(x_i, y_i), 1 \leq i \leq n$.

```
[113]: plt.plot([1, 2, 2.5, 4], [10, 12, -2, 1]);
```

[113]: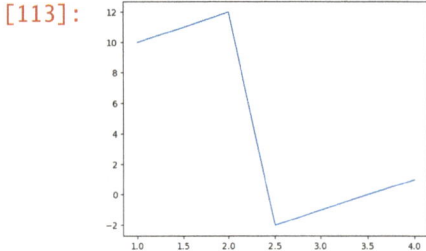

Exercise 2.5 *Plot the graph of* $\sin(x)$ *for* $0 \leq x \leq 3$.

Solution

Later we will see how to efficiently generate lists containing the data we need. Here we have to do this by "hand". We define two lists, one for the x-coordinate and the values of sin for the y-coordinate.

```
[114]: from math import sin

       x = [0, 0.3, 0.5, 0.8, 1, 1.3, 1.5, 1.8, 2, 2.3, 2.5, 2.8, 3,↵
       ↪3.3, 3.5]
       y = [sin(0), sin(0.3), sin(0.5), sin(0.8), sin(1), sin(1.3),
```

```
        sin(1.5), sin(1.8), sin(2), sin(2.3), sin(2.5), sin(2.
    ↪8),
        sin(3), sin(3.3), sin(3.5)]
```

[115]: `plt.plot(x, y);`

[115]: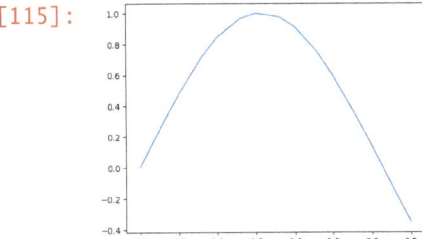

Now that we have access to the lists **x** and **y**, we can do some experiments.

[116]: ```
plt.plot(x, y[: : -1]);
plt.plot(x, y);
```

[116]: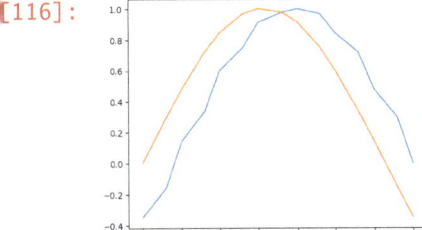

## 2.4 A First Glimpse of Importing Data

Despite the efforts of many throughout history, Isaac Newton included, it is not possible to convert cheap metals into gold. But everyone now knows that it is possible to turn data into gold, and this particular type of alchemy is called Data Science.

Python provides several libraries to import data into the computer, in order to model them, visualise them, study them and predict future behaviour. One way to import data into Python is in the form of a list. We demonstrate it here by using the library csv. More professional libraries such as panda are also available to handle data of various formats.
```

The following csv file was downloaded from the Reserve Bank of Australia home-page. It provides the interest rates since October 2002. The code shows how to use the csv library to upload the data into Python.

```
[117]: import csv

       with open('RBAdata.csv', newline='') as interest_data:
           reader = csv.reader(interest_data)
           RBA_data = list(reader)

       RBA_data[ : 10]
```

```
[117]: [['', ''],
        ['Oct-2002', '4.75'],
        ['Nov-2002', '4.75'],
        ['Dec-2002', '4.75'],
        ['Jan-2003', '4.75'],
        ['Feb-2003', '4.75'],
        ['Mar-2003', '4.75'],
        ['Apr-2003', '4.75'],
        ['May-2003', '4.75'],
        ['Jun-2003', '4.75']]
```

Once we have imported the data, we can visualise them, for example, using matplotlib. As always the data we receive is *raw* and one needs to work on it to make it ready for the process. We do this here for this example, although the method of list comprehension used will be studied in Chapter 5.

```
[118]: d = [float(item[1][0]) for item in RBA_data[1 : ]]
```

```
[119]: import matplotlib.pyplot as plt
```

```
[120]: plt.plot(d);
```

[120]:

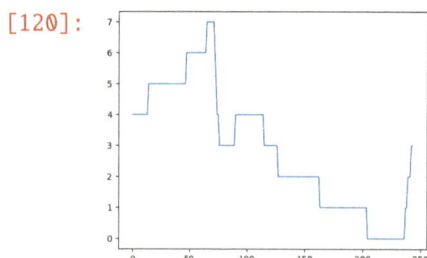

2.5 Tuples

tuple, similar to list, is designed to handle data. tuple behaves very much like list. The main difference is that one can modify an exiting list, i.e., they are *mutable objects*, whereas once a tuple is defined, one cannot modify it, i.e., they are *unmutable objects*.

The examples below show how one can define and handle tuples.

```
[121]: t = (1, 2 , 'new', 1.5)
```

Note that for a list the objects are gathered together with brackets [ ], whereas for a tuple, one uses curly brackets (). But accessing the objects within tuples is exactly like lists.

```
[122]: t[2]
```

```
[122]: 'new'
```

```
[123]: t[ : 3]
```

```
[123]: (1, 2, 'new')
```

```
[124]: t[ : : -2]
```

```
[124]: (1.5, 2)
```

The following example demonstrates what we mean when we say list is mutable whereas tuple is immutable.

```
[125]: s = [1, 2, 3]
```

```
[126]: t = (1, 2, 3)
```

```
[127]: type(s)
```

```
[127]: list
```

```
[128]: type(t)
```

```
[128]: tuple
```

```
[129]: s[1] = 'new data'
```

```
[130]: s
```

`[130]:` [1, 'new data', 3]

However executing t[1] = 'new data' generates an error. The error indicates that once a `tuple` has been defined, one cannot change it.

Tuples are defined as a collection of data between round brackets (,). In fact, one does not need to add the brackets when defining tuples.

`[131]:` seasons = 'summer' , 'autumn' , 'winter', 'spring'

`[132]:` seasons

`[132]:` ('summer', 'autumn', 'winter', 'spring')

`[133]:` good_time = seasons[0], seasons[1]

`[134]:` good_time

`[134]:` ('summer', 'autumn')

In Chapter 5 we will introduce `dictionaries` and `sets`, which are also designed to handle data, and gives us yet more power to collect, process and model data.

2.6 More Examples of Working with Data

2.6.1 Languages

The library `nltk` (Natural Language Toolkit) is a wonderful library that we can use to work with languages. NLTK has been used in many areas, including natural language processing, computational linguistics, artificial intelligence, information retrieval, and machine learning.

One needs to import and download the library and its data before one can use it.

```
[135]: import nltk
nltk.download()
```

showing info https://raw.githubusercontent.com/nltk/nltk_data/
    gh-pages/index.xml

`[135]:` True

The library `nltk` contains a variety of texts, such as the list of all English language words, or a selection of books from Project Gutenberg.

Now we can import all the English words (available in this package).

```
[136]:  from nltk.corpus import words
        word_list = words.words()
```

```
[137]:  type(word_list)
```

[137]: list

word_list is a list of over 200 thousand English words, now available to us.

```
[138]:  len(word_list)
```

[138]: 236736

```
[139]:  'fortunate' in word_list
```

[139]: True

Here is a sample list of words from word_list.

```
[140]:  word_list[ : : 20001]
```

```
[140]:  ['A',
         'beefhead',
         'commerceless',
         'Einsteinian',
         'grievingly',
         'jheel',
         'mountaintop',
         'pasilaly',
         'pun',
         'sheikly',
         'tenorite',
         'unomnipotent']
```

Here is the list of books available from Project Gutenberg.

```
[141]:  import nltk
        from nltk.corpus import gutenberg
        gutenberg.fileids()
```

```
[141]:  ['austen-emma.txt',
         'austen-persuasion.txt',
         'austen-sense.txt',
         'bible-kjv.txt',
         'blake-poems.txt',
         'bryant-stories.txt',
```

```
'burgess-busterbrown.txt',
'carroll-alice.txt',
'chesterton-ball.txt',
'chesterton-brown.txt',
'chesterton-thursday.txt',
'edgeworth-parents.txt',
'melville-moby_dick.txt',
'milton-paradise.txt',
'shakespeare-caesar.txt',
'shakespeare-hamlet.txt',
'shakespeare-macbeth.txt',
'whitman-leaves.txt']
```

We now import Hamlet into the program using the function sents, which divides the text up into its sentences, where each sentence is a list of words.

[142]: `Hamlet_sentences = gutenberg.sents('shakespeare-hamlet.txt')`

Hamlet consists of over 3000 sentences and over 37000 words.

[143]: `len(Hamlet_sentences)`

[143]: 3106

[144]: `Hamlet_sentences[1226]`

[144]: `['And', 'all', 'for', 'nothing', '?']`

[145]: `Hamlet_words = gutenberg.words('shakespeare-hamlet.txt')`

[146]: `len(Hamlet_words)`

[146]: 37360

[147]: `'Lust' in Hamlet_words`

[147]: True

[148]: `Hamlet_words.count('Lust')`

[148]: 2

We note that the contents we import via nltk are not lists, but they behave like lists as we have seen.

[149]: `type(Hamlet_words)`

```
[149]: nltk.corpus.reader.util.StreamBackedCorpusView
```

There are also some books available in nltk which we can upload.

```
[150]: import nltk
       from nltk.book import *
```

```
*** Introductory Examples for the NLTK Book ***
Loading text1, ..., text9 and sent1, ..., sent9
Type the name of the text or sentence to view it.
Type: 'texts()' or 'sents()' to list the materials.
text1: Moby Dick by Herman Melville 1851
text2: Sense and Sensibility by Jane Austen 1811
text3: The Book of Genesis
text4: Inaugural Address Corpus
text5: Chat Corpus
text6: Monty Python and the Holy Grail
text7: Wall Street Journal
text8: Personals Corpus
text9: The Man Who Was Thursday by G . K . Chesterton 1908
```

```
[151]: text3[0 : 11]
```

```
[151]: ['In',
        'the',
        'beginning',
        'God',
        'created',
        'the',
        'heaven',
        'and',
        'the',
        'earth',
        '.']
```

```
[152]: len(text3)
```

```
[152]: 44764
```

```
[153]: text3.count('God')
```

```
[153]: 231
```

We can also determine the location of a word in the text, i.e. its distance from the beginning of the text, measured in words. This positional information can be displayed using a dispersion plot. Each stripe represents an instance of a word, and each row represents the entire text

```
[154]:  text3.dispersion_plot(['God', 'earth', 'stars', 'Adam',⌴
        ⌴'Eve'])
```

[154]:

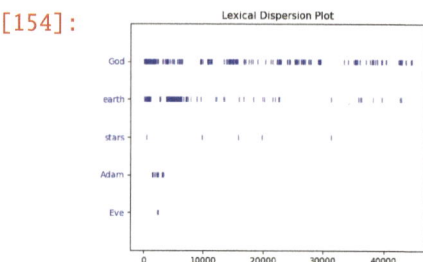

In later chapters we will be using the nltk library to write more interesting codes.

Problems

1) Define a list of the form
   z=[[x1, y1], [x2, y2], [x3, y3], [x4, y4] , [x5, y5]]
   Use Python's list methods to construct the following lists from z:
   - [x1, x2, x3, x4, x5]
   - [y1, y2, y3, y4, y5]
   - [[y1, x1], [y2, x2], [y3, x3], [y4, x4] , [y5, x5]]
   - [x1, y1, x2, y2, x3, y3, x4, y4, x5, y5]
   - [[x4, y4], [x2, y2]]
   - [[x1, y1], [x3, y3], [x5, y5]]

2) Define the following matrix in form of a list.

$$M = \begin{pmatrix} a & b & c \\ d & e & f \\ g & h & k \end{pmatrix}$$

Using list methods produce the following matrices from the list M:

- The transpose of M, i.e., $\begin{pmatrix} a & d & g \\ b & e & h \\ c & f & k \end{pmatrix}$.

- The main diagonal of M, namely $\begin{pmatrix} a & 0 & 0 \\ 0 & e & 0 \\ 0 & 0 & k \end{pmatrix}$.

- The anti-diagonal of M, namely $\begin{pmatrix} 0 & 0 & c \\ 0 & e & 0 \\ g & 0 & 0 \end{pmatrix}$.

- The matrix $\begin{pmatrix} c & b & a \\ f & e & d \\ g & h & k \end{pmatrix}$.

- The matrix $\begin{pmatrix} 1 & 2 & a \\ 3 & 4 & d \\ g & h & k \end{pmatrix}$.

- The matrix $\begin{pmatrix} \star & \star & \star \\ \star & X & \star \\ \star & \star & \star \end{pmatrix}$.

3) Consider the process

$$\begin{pmatrix} a & b & c \\ d & e & f \\ g & h & k \end{pmatrix} \longrightarrow \begin{pmatrix} a^1 & b^2 & c^3 \\ d^1 & e^2 & f^3 \\ g^1 & h^2 & k^3 \end{pmatrix}.$$

Write a code to apply this process to a 3×3-matrix. Try the code for

$$\begin{pmatrix} 1 & 2 & 3 \\ 1 & 2 & 3 \\ 1 & 2 & 3 \end{pmatrix}.$$

4) From the Reserve Bank of Australia, import data in the csv format of the interest rates since October 2002 and try to obtain various data out of the imported file (see §2.4).

Chapter 3
Decisions and Repetitions

3.1 Decision Making: The Fork on the Road

The statement

```
if cond:
    this
else:
    that
```

where cond is a Boolean expression, i.e., has the value of True or False, will execute the code block this if the cond value is True and that otherwise. That means, in either case, one of the statements this or that will be performed (but not both). So this gives us the ability to make a decision about which part of the code one wants to perform. Here is an example:

```
[1]: if True:
        print('100%')
        print('checked')
    else:
        print('0%')
```

```
[1]: 100%
    checked
```

We need to go through the above code carefully. The decoration and spacing above is not for nicety. Python uses spacing to determine the structures. The code block of this belonging to the **if** statement is determined by adding spacing, and likewise for the code block that which belongs to the **else** part of the command. In other languages one uses \begin, \end, (and) or { and } to group things together and give structure to codes, but the spacing in Python both plays this role and makes the code much easier to read. In particular, this approach cuts the number of brackets

© The Author(s), under exclusive license to Springer Nature Switzerland AG 2023
R. Hazrat, *A Course in Python*, Springer Undergraduate Mathematics Series,
https://doi.org/10.1007/978-3-031-49780-3_3

one uses in Python to group things together and it makes the code visually much more appealing. Later on in this chapter we will look at loops and repeating a block of code. That will also done by determining the code that we would like to repeat via indenting. Putting these together, a code in Python will look like the following, where the grey blocks represent structured codes.

if cond:

else:

loop

        if cond:

        else:

When working with commands that require structures, Python expects us to follow the proper indenting, as the following shows. If we run the code below, we encounter an error message, which tells us that the indenting has not been done correctly.

```
[2]: if True:
     print("100%")
     print('checked')
     else:
         print("0%")
```

```
[2]:
     Cell In[2], line 2
       print("100%")
       ^
     IndentationError: expected an indented block after 'if'
         statement on line 1
```

For the next example we look at the mysterious Collatz function, which is defined as follows:

$$f(x) = \begin{cases} x/2 & \text{if } x \text{ is even,} \\ 3x + 1 & \text{if } x \text{ is odd.} \end{cases}$$

It was conjectured by L. Collatz in 1937 that if one applies f repeatedly to any positive integer, one eventually arrives at 1, i.e., for any $x \in \mathbb{N}$, there is an n such that $f^n(x) = 1$.

We later discuss how to define functions in Python. Here we write the Collatz conditional function.

```
[3]: n=int(input('enter a number '))

     if n % 2 == 0:
         f = n//2
     else:
         f = 3 * n + 1
     f
```

enter a number 71

[3]: 214

In fact, using the method divmod we can write a rather smarter code.

```
[4]: help(divmod)
```

[4]: Help on built-in function divmod in module builtins:

     divmod(x, y, /)
         Return the tuple (x//y, x%y). Invariant: div*y + mod == x.

```
[5]: n=int(input('enter a number '))

     da = divmod(n, 2)
     if da[1] == 0:
         f = da[0]
     else:
         f = 3 * n +1
     f
```

enter a number 71

[5]: 214

Exercise 3.1 *Write a code to accept two numbers x and y and, if the difference is less than 0.00001, then print x=y.*

Solution

This is a good exercise when working with floats. As we saw in Chapter 1, when working with floats we often don't get exact values but some approximations of the computations.

```
[6]: x = 12.22; y = 12.22000002

     if abs(x-y) < 0.000001:
```

```
        print(x, "=", y)
    else:
        print(x, "not =", y)
```

[6]: 12.22 = 12.22000002

3.2 Decision Making: The Forks on the Road

So far, via **if-else**, we have been able to control the flow of the code if there are two possibilities (i.e., a fork in the road). The **if** statement is designed to control the flow of the code if there are several possibilities at hand, via the format

```
if cond:
        fork1
elif:
        fork2
elif:
        fork3
...
else:
        otherwise
```

We will define the function below, which shows how **if-elif-else** works

$$f(x) = \begin{cases} -x, & \text{if } |x| < 1 \\ \sin(x), & \text{if } 1 \leq |x| < 2 \\ \cos(x), & \text{otherwise.} \end{cases}$$

[7]:
```
import math

x = float(input('enter a number '))

if abs(x)<1:
    f = -x
elif 1<= abs(x) <2:
    f = math.sin(x)
else:
    f = math.cos(x)
f
```

enter a number 1.3098

[7]: 0.9661333622828724

Exercise 3.2 *Define the function*

$$f(x, y) = \begin{cases} e^x & \text{if } x = y \\ \frac{e^x - e^y}{x - y} & \text{if } x \neq y \end{cases}$$

and observe that for arbitrary real numbers a, b such that a < b < 0 we have

$$\frac{f(x, b)}{f(x, a)} > \frac{1 + e^b}{1 + e^a}$$

for any a ≤ x ≤ b.

Solution

We need to define the function $f(x, y)$ and then evaluate $f(x, a)$ and $f(x, b)$. It makes sense to define a stand alone function f which accepts x and y and compute $f(x, y)$. This can be done when we learn how to define functions in Python in Chapter 4. At this stage we define this function within our code.

```
[8]:  from math import exp

      print('Enter numbers a, x, and b where a<= x <= b and a<b<0')
      a = float(input('enter a: '))
      b = float(input('enter b: '))
      x = float(input('enter x: '))

      y = a
      if x == y:
          fa = exp(x)
      else:
          fa = (exp(x) - exp(y))/(x - y)
      y = b
      if x == y:
          fb = exp(x)
      else:
          fb = (exp(x) - exp(y))/(x - y)

      z = (1 + exp(b))/(1 + exp(a))
      print(fb / fa > z)
```

```
Enter numbers a, x, and b where a<= x <= b and a<b<0
enter a: -3.4
enter b: -1.44
enter x: -2
```

[8]: True

Exercise 3.3 *Write a code to check if a number is an integer or a float, and whether it is positive or negative.*

Solution

Recall that the function `type` will give back the type of the object. Thus using `type` we can determine if the input is an integer or a float.

```
[9]: x=-3.14
     if type(x) != int and type(x) != float:
         print('input is not a number')
     elif x == 0:
         print(x, 'is zero')
     elif x > 0:
         if type(x) == int:
             print(x, 'is a positive integer')
         else:
             print(x, 'is a positive real number')
     else:
         if type(x) == int:
             print(x, 'is a negative integer')
         else:
             print(x, 'is a negative real number')
```

[9]: -3.14 is a negative real number

Note in the code above how an **if-else** is used within another conditional structure.

Exercise 3.4 *Write a code to accept a string and determine if it is palindromic.*

Solution

Recall we can reverse a string by reading its characters from right to left as follows:

```
[10]: x = 'university'
      x[: : -1]
```

[10]: 'ytisrevinu'

```
[11]: x = input('Enter a word: ')

      if x == x[ : : -1]:
          print(x, "is palindromic")
```

```
else:
    print(x, "is not palindromic")
```

Enter a word: kayak

[11]: kayak is palindromic

Exercise 3.5 *Write a code to accept a list and then remove the first and last element of the list if they are the same. Enhance the code so that if the list is empty it says the list is palindromic. Further enhance the code so that it works for strings as well.*

Solution

Let us start with a list, capture the first and last elements and then print all other elements. We have all the tools to do this.

```
[12]: x = ['a', 'b', 'c', 'd', 'e', 'a']
      print(x[0], x[-1], ' and the list without the first and last
        ↪element ', x[1 : -1])
```

[12]: a a and the list without the first and last element
         ['b', 'c', 'd', 'e']

This shows how we should proceed.

```
[13]: if x[0] == x[-1]:
          x = x[1 : -1]
      else:
          x
```

```
[14]: x
```

[14]: ['b', 'c', 'd', 'e']

Next we check if the list is empty to start with.

```
[15]: if x == []:
          print('palindromic')
      elif x[0] == x[-1]:
          x = x[1 : -1]
          print(x)
      else:
          print(x)
```

[15]: ['b', 'c', 'd', 'e']

Let us check the code with x=['a','yes','c','x','c','yes','a'].

```
[16]: x = ['a', 'yes', 'c', 'x', 'c', 'yes', 'a']
```

You will see if we run the code 4 times, we get the word 'palindromic'. This shows a way to write a code for checking if a list is palindromic, and we can make it more efficient if we have tools to repeat a block of codes. This is done next when we introduce *loops* in Python.

It is very easy to modify the code to handle strings as well.

```
[17]: x = 'kayak'

if x == [] or x == '':
    print("Palindromic")
elif x[0] == x[-1]:
    x = x[1 : -1]
    print(x)
else:
    print(x)
```

```
[17]: aya
```

3.3 Loops and Repetitions

3.3.1 For-loop

A primary ability that a computer language provides is the ability to repeat a certain code 'fast'. Python provides three kind of *loops* that enable us to repeat part of our codes. These are quite similar to the loops that exist in any procedural language like Fortran or C. We will introduce **for** and **while** loops and study nested loops, that is, loops defined inside each other. We finish the chapter by looking at other nested commands.

The first way to create a repetition is to define a list and ask Python to go through the list. This way we can repeat a block as the number of elements of the list.

```
for i in list:
        block
```

The block itself could consists of a large number of lines. Similar to an if statement, we specify the block belonging to the **for**-loop with an appropriate spacing. The program goes through the list and each time runs the block.

```
[18]: for n in [2, 3, 4, 5]:
    print(n)
```

```
[18]:  2
       3
       4
       5
```

Here the parameter n goes through the list (from left to right) [2, 3, 4, 5] each time picking up one of the items in the list and executing the block, here the line print(n). The program terminates when n reaches the end of the list.

As one can see, the arrangement is very similar to an if statement.

```
[19]:  n = 4
       if n in [2,3,4,5]:
           print(n)
```

```
[19]:  4
```

Going through a list to create a loop gives us an amazing power, as we can create loops going through practically all kind of lists which contain all kinds of objects!

```
[20]:  my_list=['Western', 'Sydney', 'University']
```

```
[21]:  for item in my_list:
           print(item + ' yeah!')
```

```
[21]:  Western yeah!
       Sydney yeah!
       University yeah!
```

```
[22]:  for name in ['western', 'sydney', 'university']:
           print(name.upper())
```

```
[22]:  WESTERN
       SYDNEY
       UNIVERSITY
```

Python provides a range object which comes in very handy: one can create a loop using range. range(m) creates a range from 0 to $m - 1$.

```
[23]:  for i in range(7):
           print('phishing' * i)
```

```
[23]:  phishing
       phishingphishing
       phishingphishingphishing
       phishingphishingphishingphishing
```

phishingphishingphishingphishingphishing
phishingphishingphishingphishingphishingphishing

```
[24]: for i in range(1, 15, 5):
          print('phishing' * i)
```

[24]: phishing
      phishingphishingphishingphishingphishingphishing
      phishingphishingphishingphishingphishingphishingphishingphishi
          ngphishingphishingphishing

Note that range is neither a list nor a tuple, it is just an *iterator* object. As promised, we will create a list containing images and then create a loop going through the images and modify them accordingly.

```
[25]: from PIL import Image
      x = Image.open('dog.jpg')
      y = Image.open('Napolean.jpg')
```

```
[26]: for pic in [x, y]:
          display(pic)
```

[26]:

```
[27]:  scale = 1
       for pic in [x, y] * 2:
           s = (pic.size[0] // scale, pic.size[1] // scale)
           display(pic.resize(s))
           scale *= 2
```

[27]:

Exercise 3.6 *A word is called palindromic if it reads the same backwards as forwards, e.g., madam or kayak. Using the library* nltk *import all the words in the English language into Python and then find all the palindromic words.*

Solution

Recall that the library nltk provides tools and data related to language.

```
[28]: from nltk.corpus import words
      word_list = words.words()
```

Now finding the palindromic words is an easy test going through the whole list.

```
[29]: for word in word_list:
          if word == word[ : : -1] and len(word) > 3:
              print(word, end=' ')
```

```
[29]: acca adda affa ajaja alala alula amma anana anna arara atta
      boob civic deed deedeed degged elle hallah immi kakkak kayak
      keek kelek lemel level maam madam mesem minim murdrum noon
      otto peep poop radar redder refer repaper retter rever reviver
      rotator rotor siris sooloos tebbet teet tenet terret toot
      ululu yaray
```

Exercise 3.7 *Find all palindromic words in the book of Genesis.*

Solution

This exercise is quite similar to the previous one.

```
[30]: import nltk
      from nltk.book import *
```

```
[30]: *** Introductory Examples for the NLTK Book ***
      Loading text1, ..., text9 and sent1, ..., sent9
      Type the name of the text or sentence to view it.
      Type: 'texts()' or 'sents()' to list the materials.
      text1: Moby Dick by Herman Melville 1851
```

```
text2: Sense and Sensibility by Jane Austen 1811
text3: The Book of Genesis
text4: Inaugural Address Corpus
text5: Chat Corpus
text6: Monty Python and the Holy Grail
text7: Wall Street Journal
text8: Personals Corpus
text9: The Man Who Was Thursday by G . K . Chesterton 1908
```

[31]:
```python
for word in text3:
    if len(word)> 3 and word == word [ : : -1] :
        print(word)
```

[31]: noon
       noon
       deed

[32]:
```python
texts()
```

[32]: text1: Moby Dick by Herman Melville 1851
       text2: Sense and Sensibility by Jane Austen 1811
       text3: The Book of Genesis
       text4: Inaugural Address Corpus
       text5: Chat Corpus
       text6: Monty Python and the Holy Grail
       text7: Wall Street Journal
       text8: Personals Corpus
       text9: The Man Who Was Thursday by G . K . Chesterton 1908

Exercise 3.8 *The sum of two positive integers is* 5432 *and their least common multiple is* 223020. *Find the numbers.*

Solution

The math library does not yet include the function for the least common multiple of two numbers. However, the greatest common divisor is available via math.gcd. On the other hand, we know the identity

$$\text{lcm}(n, m) = \frac{mn}{\gcd(n, m)}.$$

This allows us to write the code for the exercise.

[33]:
```python
import math
math.gcd(25,30)
```

[33]: 5

```
[34]: (25 * 30) // math.gcd(25,30)
```

[34]: 150

```
[35]: import math

for n in range(1, 5433):
    m = 5432 - n
    lcm = (n * m) // math.gcd(n,m)
    if lcm == 223020:
        print(n, m)
```

[35]: 1652 3780
      3780 1652

As there is a symmetry between n and m we need to go up to half of 5432 in order to avoid repetition of results.

```
[36]: import math

for n in range(1, 5433//2):
    m = 5432 - n
    lcm = (n * m) // math.gcd(n,m)
    if lcm == 223020:
        print(n, m)
```

[36]: 1652 3780

Exercise 3.9 *Determine all the positive integers n between 3 and 50 for which* 2^{2008} *is divisible by*

$$1 + \binom{n}{1} + \binom{n}{2} + \binom{n}{3}.$$

Here $\binom{n}{m}$ *is the binomial coefficient:* $\binom{n}{m}$ *is defined as* $\dfrac{n!}{m!(n-m)!}$

Solution

Both factorial and binomial functions are available in the math library, as the following examples show:

```
[37]: import math
      math.factorial(5) == 1 * 2 * 3 * 4 * 5
```

```
[37]: True
```

```
[38]: math.comb(5,3) == math.factorial(5)//(math.factorial(3) *
      ↪math.factorial(5-3))
```

```
[38]: True
```

We are ready to write the code. The difficulty is in translating the large formula into Python correctly.

```
[39]: for n in range(3,51):
          f = (1 + math.comb(n,1) +  math.comb(n,2) +  math.
      ↪comb(n,3))
          if (2**2008) % f== 0:
              print(n)
```

```
[39]: 3
      7
      23
```

Please appreciate the complexity of the computation!

```
[40]: 2**2008
```

```
[40]: 293921457990209158203605299501486587909713331734705971322276540
      627396162916446800347304828497025605099122166947580790470002462
      453980942164845038427178663215460172772211999436801763274619494
      514870858053094562524786640935586934754211705131586663593866165
      516791188895740950898251790395677822812580408244051664241072407
      000213774342091481108259990786393027841098246954768962126136340
      818524880106908845781292048893428214830405175756437514347929224
      149123944676950789355316620691925989560420249809810474574291853
      773889494338599752572893233746059542823106006739520449114953730
      10647749329399156163119321894151520256
```

Exercise 3.10 *Notice that* $12^2 = 144$ *and* $21^2 = 441$, *i.e., the numbers and their squares are reverses of each other. Find all the numbers up to 10000 with this property.*

Solution

The first thing to take care of is the reversal of the digits of a number. We can use the trick that we used for strings, reading the characters of a string from right to left via s[: : -1]. So we convert a number into a string, read the characters from right to left, and then convert it back to a number.

```
[41]: x=str(12345)
```

```
[42]: x[ : : -1]
```

[42]: '54321'

```
[43]: int(x[ : : -1])
```

[43]: 54321

```
[44]: int(str(12345)[ : : -1])
```

[44]: 54321

Now we need to translate the property we are after into code, that is, we need to raise a number to the power of two, reverse it, and then compare the result to its reverse to the power of two; in code it reads:

int(str(n**2)[: : -1]) == int(str(n)[: : -1])**2.

Putting these together:

```
[45]: for n in range(1,10000):
          if int(str(n**2)[ : : -1]) == int(str(n)[: : -1])**2:
              print(n, end=' ')
```

```
[45]: 1   2   3   10   11   12   13   20   21   22   30   31   100   101   102
      103   110   111   112 113   120   121   122   130   200   201   202
      210   211   212   220   221   300   301   310 311   1000   1001   1002
      1003   1010   1011   1012   1013   1020   1021   1022   1030 1031
      1100   1101   1102   1103   1110   1111   1112   1113   1120   1121
      1122   1130 1200   1201   1202   1210   1211   1212   1220   1300
      1301   2000   2001   2002   2010 2011   2012   2020   2021   2022
      2100   2101   2102   2110   2111   2120   2121   2200 2201   2202
      2210   2211   3000   3001   3010   3011   3100   3101   3110   3111
```

We check one of the answers:

```
[46]: 112**2
```

[46]: 12544

```
[47]: 211**2
```

[47]: 44521

We revisit Exercise 1.1 on cyclic numbers.

Exercise 3.11 *A number with n digits is called* **cyclic** *if multiplication by* $1, 2, 3, \cdots, n$ *produces numbers with the same digits in a different order. Find the only 6-digit cyclic number.*

Solution

Let's spill the beans: The number 142857 is the only 6-digit cyclic number! We first check this via a simple loop.

```
[48]: for i in range(1, 7):
          print(i * 142857, end=' ')
```

[48]: 142857 285714 428571 571428 714285 857142

How to actually find this number? One approach is to sort all the digits of the number from smaller to larger. Then we check that any multiple of the number, when similarly sorted, has the same sequence of digits.

```
[49]: str(142857)
```

[49]: '142857'

```
[50]: sorted(str(142857))
```

[50]: ['1', '2', '4', '5', '7', '8']

```
[51]: str(142857 * 2)
```

[51]: '285714'

```
[52]: sorted(str(142857)) == sorted(str(142857 * 2))
```

[52]: True

Here is the naive way to code this approach:

```
[53]: for i in range(100000, 1000000):
          if sorted(str(i)) == sorted(str(i * 2)) == sorted(str(i
      ↪* 3)) == sorted(str(i * 4))== sorted(str(i * 5)) ==↵
      ↪sorted(str(i * 6)):
              print(i)
```

[53]: 142857

Here is a bit of a re-working of the above code. We set x to be the digits of a number, sorted, and a boolean value y = True. Then we examine multiples of the number, each time comparing the result with x. If the arrangements do not match, we change the value of the Boolean statement to false. This will keep track of whether the arrangement of the digits changes.

```
[54]:  x = sorted(str(142857))
       y = True
       for i in range(1, 7):
           y = y and x == sorted(str(142857 * i))
       print(y)
```

[54]: True

Recall that, for some operation #, we can write y = y # s in the shorter form y #= s.

```
[55]:  x = sorted(str(142857))
       y = True
       for i in range(1, 7):
           y &= x == sorted(str(142857 * i))
       print(y)
```

[55]: True

We can improve the above code even more. If in some instance the value of y becomes false, namely, if the digits of the new number differs from the digits of x, we do not need to go further through the rest of the loop. In this case, we can stop the loop, using the command break. The break statement breaks out of the loop entirely.

```
[56]:  x = sorted(str(142857))
       y == True
       for i in range(1, 7):
           if x != sorted(str(142857 * i)):
               y = False
               print(y)
               break
```

Now we incorporate the code, running x through all six-digit numbers.

```
[57]:  for i in range(100000, 1000000):
           x = sorted(str(i))
           y = True
```

```
    for k in range(1, 7):
        if x != sorted(str(i * k)):
         y=False
         break
    if y: print(i)
```

`[57]:` 142857

Exercise 3.12 *Define* $f(x) = \sqrt{1+x}$. *We have*

$$f(f(f(f(f(x))))) = \sqrt{1 + \sqrt{1 + \sqrt{1 + \sqrt{1 + \sqrt{1+x}}}}}.$$

For any give x, calculate the above expression.

Solution

Later we will define the function f in Python and directly calculate the composition of functions. At the moment the only tool we have is loops.

`[58]:`
```
import math
x = 0
for _ in range(5):
    x = math.sqrt(1 + x)
print(x)
```

`[58]:` 1.6118477541252516

This is the first time we have used _ as a variable within a loop. If one does not explicitly need the variable which runs through the loop, one can use _ instead.

Exercise 3.13 *Find n between 1 to 100 such that* $2^n - 1$ *is divisible by 7.*

Solution

We create a loop, let n go through the list from 1 to 100, and on each occasion check if $2^n - 1$ is divisible by 7. This is done by checking if the remainder of $2^n - 1$ by 7 is zero: in Python: `if (2**n - 1) % 7 == 0`.

`[59]:`
```
for n in range(1, 101):
    if (2**n - 1) % 7 == 0:
        print(n, end=' ')
```

[59]: 3 6 9 12 15 18 21 24 27 30 33 36 39 42 45 48 51 54 57 60 63
       66 69 72 75 78 81 84 87 90 93 96 99

Exercise 3.14 *Plot the function* $f(x) = \sin(x)$, *for* $0 \leq x \leq \pi$.

Solution

Recall that the library `matplotlib` is used for plotting data. We need to create two
lists x and y then `plt.plot(x,y)` will produce a graph determined by pairs from
the lists. This time we can use the for-loop to generate the lists.

[60]:
```python
import math, matplotlib.pyplot as plt

x = []
y = []
for i in range(0, 200):
    step = i * 2 * math.pi / 200
    x.append(step)
    y.append(math.sin(step))

plt.plot(x,y)
```

[60]:
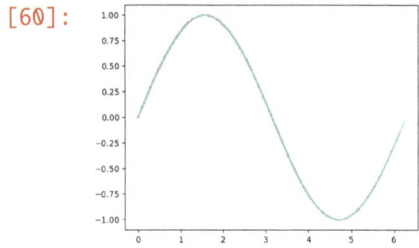

Exercise 3.15 *Consider the complex function (i is the imaginary number here)*

$$g(t) = e^{-2it} + \frac{1}{2}e^{5it} + \frac{1}{5}e^{19it}$$

and plot the graph where x and y are the real and imaginary part of $g(t)$ *for*
$0 \leq t \leq 2\pi$.

Solution

We need to use the exponential function with complex numbers. The library `math`
provides the function `exp`, however here we can only use real numbers with this
function. The library `cmath` provides functions allowing us to work with complex
numbers. For this reason we import the function `exp` from the library `cmath`.

Once we calculate the function g, we can retrieve the real part with `g.real` and the imaginary part with `g.imag`. We collect these values into two lists of `x` and `y` and using `matplotlib` we plot the graph.

```
[61]: from cmath import exp
      import matplotlib.pyplot as plt

      x = []
      y = []
      for t in range(0, 2000):
          s = t * 2 * math.pi / 200
          g = exp(-2j * s) + (1/2)*exp(5j * s) + (1/5)*exp(19j * s)
          x.append(g.real)
          y.append(g.imag)

      plt.axis('off');
      plt.axis('equal');
      plt.plot(x,y);
```

[61]:

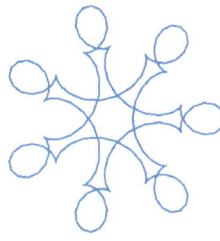

Finally, we revisit the interest rate data we have imported from the Reserve Bank of Australia. Now we have a tool to extract parts of the list.

```
[62]: import csv

      with open('RBAdata.csv', newline='') as interest_data:
          reader = csv.reader(interest_data)
          RBA_data = list(reader)

      RBA_data[ : 10]
```

```
[62]: [['', ''],
       ['Oct-2002', '4.75'],
       ['Nov-2002', '4.75'],
       ['Dec-2002', '4.75'],
       ['Jan-2003', '4.75'],
       ['Feb-2003', '4.75'],
```

```
['Mar-2003', '4.75'],
['Apr-2003', '4.75'],
['May-2003', '4.75'],
['Jun-2003', '4.75']]
```

First, we remove the empty entry at the start of the list.

```
[63]:  del RBA_data[0]
       RBA_data[ : 10]
```

```
[63]:  [['Oct-2002', '4.75'],
        ['Nov-2002', '4.75'],
        ['Dec-2002', '4.75'],
        ['Jan-2003', '4.75'],
        ['Feb-2003', '4.75'],
        ['Mar-2003', '4.75'],
        ['Apr-2003', '4.75'],
        ['May-2003', '4.75'],
        ['Jun-2003', '4.75'],
        ['Jul-2003', '4.75']]
```

```
[64]:  dates = []
       data = []
       for d in RBA_data:
           dates.append(d[0])
           data.append(float(d[1][0]))
```

```
[65]:  import matplotlib.pyplot as plt

       plt.plot(data);
```

3.3.2 Nested For-loops

In many applications there are several factors (variables) which change simultaneously, and this calls for what we call a *nested loop*. Instead of trying to describe the situation abstractly, let us look at some examples.

Let us find all the pairs (n, m) for $n, m \leq 10$ such that $n^2 + m^2$ is a square number (e.g., $(3, 4)$ as $3^2 + 4^2 = 5^2$).

Note that here we are working with two parameters n and m, and each of them has a range between 1 and 10. We can design two for-loops, each taking care of one of the variables.

```
[66]:  for n in range(1, 11):
           for m in range(1, 11):
               if math.sqrt(n**2 + m**2).is_integer():
                   print(n, " ", m)
```

```
[66]:  3    4
       4    3
       6    8
       8    6
```

Here the outer loop starts with the counter n getting the value 1. Then it is the turn of the block inside this loop, which is again another loop run. In the inner loop m in (1, 11) makes the counter m run from 1 to 10. This done, in the outer loop n takes the value 2 and then m runs from 1 to 10 and so on, each time checking whether $\sqrt{n^2 + m^2}$ is an integer. This is done with a method available in floats, namely .is_integer.

As one can see, we get the pairs $(3, 4)$ and $(4, 3)$, which for us are the same answer. In order to remove this repetition, we can modify the code.

```
[67]:  for n in range(1, 11):
           for m in range(i, 11):
               if math.sqrt(n**2 + m**2).is_integer():
                   print(n, " ", m)
```

The reader should see that in the inner loop, m needs to start from n and goes all the way to 10. This is enough to find all the pairs up to 10 with the desired property. Can you say how many times the if line is going to be performed?

Exercise 3.16 *Pick an odd number p. Then find a pair (q, r) of positive integers such that $p^2 + q^2 = r^2$.*

Solution

Clearly we have to set a bound for q, let's say we will check up to 100 for q and r. The following code should find all the pairs up to 100.

```
[68]: p= int(input('enter an odd number:'))
      for q in range(1, 101):
          for r in range(q+1, 101):
              if r**2 - q**2 == p**2:
                  print(q, ' ', r)
```

enter an odd number:33

```
[68]: 44    55
      56    65
```

Exercise 3.17 *We asked ChatGPT to write a Python code to find the smallest number that can be written as a sum of two cubes. Here is the code. What is the issue with it!?*

```
[69]:     # Initialize the smallest sum to a large value
          smallest_sum = float('inf')

          # Define the upper limit for the search (You can adjust␣
          ↪this if needed)
          upper_limit = 1000

          # Iterate through all possible combinations of cube␣
          ↪numbers
          for a in range(1, upper_limit):
              for b in range(a, upper_limit):
                  # Calculate the sum of the cubes of a and b
                  current_sum = a**3 + b**3

                  # Check if the current sum is smaller than the␣
          ↪smallest sum found so far
                  if current_sum < smallest_sum:
                      smallest_sum = current_sum

      print(f"The smallest number that can be written as a sum of␣
      ↪two cubes is: {smallest_sum}")
```

```
[68]: The smallest number that can be written as a sum of two cubes
      is: 2
```

Solution

The smallest positive integer that can be written as the sum of two positive cubes is obviously $2 = 1^3 + 1^3$. This does not require a code! However the code provided by ChatGPT finds the answer correctly (although the nested loop runs 1 million times!) and the overall arrangement of the code also is quite impressive.

Exercise 3.18 *Define a 3×2 matrix (a_{ij}) with entries $a_{ij} = i - j$. Then find the sum of all the entries.*

Solution

We have seen how to handle matrices using lists. Now that we have the ability to create loops, we can generate the entries of the matrix systematically. We first represent a generic 3×2-matrix by A=[[0,0], [0,0], [0,0]]. Next translating $a_{ij} = i - j$ into code we get A[i][j] = i - j. Our task now is to run i and j from 0 to 2.

```
[70]: A=[[0,0], [0,0], [0,0]]

      for i in range(3):
          for j in range(2):
              A[i][j] = i - j
      print(A)

      X = 0
      for i in range(3):
          for j in range(2):
              X += A[i][j]
      print(f'The sum of all the entries is {X}')
```

```
[70]: [[0, -1], [1, 0], [2, 1]]
      The sum of all the entries is 3
```

Exercise 3.19 *Generate the following matrix for different values of n.*

$$\begin{pmatrix} 1 & 2 & \cdots & n \\ n+1 & n+2 & \cdots & 2n \\ \vdots & \vdots & \vdots & \vdots \\ \cdots & \cdots & \cdots & n^2 \end{pmatrix}$$

Solution

Here is a step-by-step process describing how to create such a matrix, when $n = 5$.

```
[71]: L = []
      for i in range(5):
          L += [i+1]
      print(L)
```

[71]: [1, 2, 3, 4, 5]

```
[72]: L=[]
      for j in range(5):
          for i in range(1, 6):
              L += [i + 5*j]
      print(L)
```

[72]: [1, 2, 3, 4, 5, 6, 7, 8, 9, 10, 11, 12, 13, 14, 15, 16, 17, 18, 19, 20, 21, 22, 23, 24, 25]

```
[73]: S = []
      for j in range(5):
          L = []
          for i in range(1, 6):
              L += [i + 5*j]
          S += [L]
      print(S)
```

[73]: [[1, 2, 3, 4, 5], [6, 7, 8, 9, 10], [11, 12, 13, 14, 15], [16, 17, 18, 19, 20], [21, 22, 23, 24, 25]]

We can approach this exercise differently, by splitting a list. We consider the list of numbers 1 to 25 and each time collect 5 consecutive elements from the list and append it to a new list.

```
[74]: size = 5
      m = [*range(1, size**2+1)]
      mat = []
      for i in range(0, len(m), size):
          mat.append(m[i : i + size])
      print(mat)
```

[74]: [[1, 2, 3, 4, 5], [6, 7, 8, 9, 10], [11, 12, 13, 14, 15], [16, 17, 18, 19, 20], [21, 22, 23, 24, 25]]

Exercise 3.20 Write a code to multiply two 3×3-matrices.

Solution

Recall that if A is a $n \times m$ matrix and B is a $m \times p$ matrix, then the product matrix $C := A \cdot B$ is an $n \times p$ matrix, where the entries of C are

$$C_{i,j} = \sum_{k=1}^{m} a_{ik} b_{kj},$$

where $1 \leq i \leq n$ and $1 \leq j \leq p$. We can translate this into Python via three nested-loops. We will later see that several Python libraries such as sympy and numpy provide matrix multiplication, allowing us to perform linear algebra.

```
[75]:  A = [[2, 1, -1], [-1, 2, 0], [1,1,3]]
       B = [[-1, 0, 2], [-1, -1, 1],[-2, 0, -1]]
       C = [[0, 0, 0],[0, 0, 0],[0, 0, 0]]
```

We define two sample matrices A and B, and C is designed to collect the entries of the product of A and B.

```
[76]:  for i in range(3):
           for j in range(3):
               for k in range(3):
                   C[i][j] += A[i][k] * B[k][j]
       print(C)
```

```
[76]:  [[-1, -1, 6], [-1, -2, 0], [-8, -1, 0]]
```

Exercise 3.21 *Find the number of palindromic words in all the standard books available in* nltk.

Solution

We have already seen how to find the palindromic words in any of the given books. All we need to do is to create another loop which runs through all the books available.

```
[77]:  import nltk
       from nltk.book import *
```

```
[78]:  for book in [text1, text2, text3, text4, text5, text6, text7,↵
       ↪text8, text9]:
           sum=0
           for word in book:
               if len(word)> 3 and word == word [ : : -1]:
```

```
        sum += 1
    print(f'The number of palindromic words in {book} is␣
↪{sum}')
```

[78]: The number of palindromic words in <Text: Moby Dick by Herman
      Melville 1851> is 75
      The number of palindromic words in <Text: Sense and
      Sensibility by Jane Austen 1811> is 9
      The number of palindromic words in <Text: The Book of
      Genesis> is 3
      The number of palindromic words in <Text: Inaugural Address
      Corpus> is 22
      The number of palindromic words in <Text: Chat Corpus> is 629
      The number of palindromic words in <Text: Monty Python and
      the Holy Grail> is 5
      The number of palindromic words in <Text: Wall Street
      Journal> is 42
      The number of palindromic words in <Text: Personals Corpus>
      is 0
      The number of palindromic words in <Text: The Man Who Was
      Thursday by G . K . Chesterton 1908> is 7

3.3.3 While Loops

While loops provide another way to repeat a block of code. This time the block is
going to be repeated until a certain condition is satisfied, i.e., a boolean expression
becomes True. The while-loop has the form

```
while cond:
    block
```

We start with the following example. Consider $n = 123$; while n is not divisible by
7, append the digit 1 to the right of n. Note that we don't know from the outset how
many 1s we need to add to the right-hand side of the given number until it becomes
divisible by 7.

[79]:
```
n = 123
while n % 7 !=0:
    n = n*10 + 1
print(n)
```

[79]: 1231111

n%7 !=0 is our Boolean statement (condition). The while-loop repeats the block belongs to it (specified via spacing) while n%7 !=0 returns True. Here n=10n+1 is the block of code we want to repeat. The code n=10n+1 simply takes the number n and places 1 at the far right of the number (right?). So the aim is to put as many 1s to the right of the original n, which is 123 here, to get a number divisible by 7. The While loop does exactly this. It is going to repeat the above code until n%7 !=0 becomes False. That is, until *n* becomes divisible by 7. And this is what we were looking for.

Exercise 3.22 *Find the smallest positive integer m such that* $529^3 + 132^3 m$ *is divisible by* 262417.

Solution

```
[80]: m = 1
      while (529**3 + (132**3)*m) % 262417 !=0:
          m += 1
      m
```

[80]: 1984

We start with m=1 and while the remainder on division of 529**3+(132**33)*m by 262417 is not zero, in Python terms, while (529**3+(132**3)*m) % 262417 !=0: we add one to m, i.e., m += 1, and repeat this until the remainder is zero. Then this is the m we are looking for.

Exercise 3.23 *Find the smallest multiple of* 99999 *that contains no 9s amongst its digits.*

Solution

Recall that within a string we could check if an element belongs to it (similar to lists). So we convert the number into a string and check whether the digit 9 belongs to it.

```
[81]: n = 99999
      i = 1
      while '9' in str(n):
          i += 1
          n = 99999 * i
      print(f'the {i}th multiple of 99999 is {n} which has no digit↵
          ↪9')
```

[81]: the 11112th multiple of 99999 is 1111188888 which has no
      digit 9

Note that it is important to get the flow of the code right, if we swap the two lines in the block of the while-loop the answer is not correct.

```
[82]: n = 99999
      i = 1
      while "9" in str(n):
          n = 99999 * i
          i += 1
      print(i, "   ", n)
```

```
[82]: 11113     1111188888
```

Although the library math provides the function gcd to calculate the greatest common divisor, here we try to write an efficient way to compute the gcd of two given numbers. Let us start with a naive approach:

```
[83]: n = 36; m = 16;
      if n > m:
          small = m
      else:
          small = n
      for i in range(1, small + 1):
          if (n % i == 0) and (m % i == 0):
              gcd = i
      gcd
```

```
[83]: 4
```

The code starts with $i = 1$ and if i divides both n and m, it would be collected into the variable gcd. We then increase i and test again. The loop runs i from 1 up to the smaller number n or m, and eventually it gives us gcd. We could improve the code by starting i from the smaller number and decreasing its value, checking each time if i divides both n and m. As soon as this happens, we stop the loop via break and print this value, which would be the greatest common divisor of n and m.

```
[84]: n = 36; m = 16;
      if n > m:
          small = m
      else:
          small = n
      for i in range(small, 1 , -1):
          if (n % i == 0) and (m % i == 0):
              gcd = i
              break
      gcd
```

[84]: 4

We can use the elegant Euclidean algorithm, which says that if for integers n and m we have $n = mq + r$, where $q, r \in \mathbb{Z}$, then $\gcd(n, m) = \gcd(m, r)$. Using this fact we can write:

```
[85]:  n = 36; m = 16
       while m != 0:
           n , m = m, n % m
       print(n)
```

[85]: 4

```
[86]:  n = 2334426; m = 3336
       while m != 0:
           n, m = m, n % m
       n == math.gcd(2334426, 3336)
```

[86]: True

Exercise 3.24 *Write a code to get a number and create a list of its digits. Modify the code to give the list of digits in any given base.*

Solution

Recall that for any $n \in \mathbb{N}$, and any $1 \le b \le 10$, one can write

$$n = \sum r_i b^i$$

where $0 \le r_i < b$ are all unique. As an example

$$1234 = 2 \times 4^0 + 0 \times 4^1 + 1 \times 4^2 + 3 \times 4^2 + 0 \times 4^3 + 1 \times 4^4.$$

Then we say the number 1234 can be written as 103102 in base 4. If we choose b to be 10, then we retrieve all the digits of the number.

The code is easy to compile, if we recall that we can write n = n // b + n % b.

```
[87]:  n = int(input('enter an integer '))
       d = []
       while n !=0:
           d.append(n % 10)
           n = n // 10
       d.reverse()
       d
```

enter an integer 2345

[87]: [2, 3, 4, 5]

Now that the code is working with base 10, we can modify it to work for any base.

```
[88]: n = int(input('enter an integer '))
b = int(input('enter a base '))
d = []
while n !=0:
    d.append(n % b)
    n = n // b
d.reverse()
d
```

enter an integer 2345
enter a base 4

[88]: [2, 1, 0, 2, 2, 1]

Exercise 3.25 *Write a code to check if a number k is of the form* $2^m 3^n$*. Enhance the code to find the m and n and write* $k = 2^m 3^n$*.*

Solution

As we don't know how many 2s and 3s appear in the decomposition of the number k, we can use a while loop to keep dividing k by 2 until the result is no longer divisible by 2. Next we get the result and start dividing it by 3, until the result is no longer divisible by 3. If the result has been reduced to 1, it means the original number must be of the form $k = 2^m 3^n$.

```
[89]: k = l = 16 * 3 * 2 * 5
while l % 2 == 0:
    l = l // 2
while l % 3 == 0:
    l = l // 3
if l == 1:
    print(f'{k} is of the form 2^m 3^n')
else:
    print(f'{k} is not of the form 2^m 3^n')
```

[89]: 480 is not of the form 2^m 3^n

Next we modify the code slightly and keep track of the number of times we divide the number by 2 and 3, respectively.

```
[90]:  k = 1 = 2**5 * 3**7
       sum2 = 0; sum3 = 0
       while l % 2 == 0:
           l = l // 2
           sum2 += 1
       while l % 3 == 0:
           l = l // 3
           sum3 += 1
       if l == 1:
           print(f'{k} is of the form 2^{sum2} 3^{sum3}')
       else:
           print(f'{k} is not of the form 2^m 3^n')
```

```
[90]:  69984 is of the form 2^5 3^7
```

Problems

1) Find the number of positive integers $0 < n < 20000$ such that 1997 divides $n^2 + (n + 1)^2$. Try the same code for 2009 and 2022.

2) Show that the number of k between 0 and 1000 for which $\binom{1000}{k}$ is odd is a power of 2.

   Note that $\binom{n}{m}$ is the binomial coefficient defined by

   $$\binom{n}{m} = \frac{n!}{m!(n-m)!}$$

   and it is available in the Python math library as the command comb(n,m).

3) For integers $2 \le n \le 200$, find all n such that n divides $(n - 1)! + 1$. Show that there are 46 such n.

4) Show that there is only one positive integer n smaller than 100 such that $n! + (n + 1)!$ is the square of an integer.

5) Let m be a natural number and

   $$A = \frac{(m + 3)^3 + 1}{3m}.$$

   Find all the integers m less than 500 such that A is an integer. Show that A is always odd.

6) For a given n, calculate the series

$$1^3 + 2^3 + \cdots + n^3,$$

$$1 + \frac{1}{1} + \frac{1}{2!} + \cdots + \frac{1}{n!}$$

7) Show that the only n less than 1000 such that

$$3^n + 4^n + \cdots + (n+2)^n = (n+3)^n$$

   are the numbers 2 and 3.

8) Find all the numbers up to one million which have the following property: if $n = d_1 d_2 \cdots d_k$ then $n = d_1! + d_2! + \cdots + d_k!$ (e.g. $145 = 1! + 4! + 5!$).

9) Consider the number 485. Observe that $485 + 584 = 1069$ is a prime number. Find all numbers n between 1 and 1000 such that n plus its reverse is prime.

10) Write two functions f and g, each accepting two sequences $(x_1, x_2, \ldots, x_n)$ and $(y_1, y_2, \ldots, y_n)$ of positive integers, and return, respectively,

$$\sqrt{x_1 y_1} + \sqrt{x_2 y_2} \cdots + \sqrt{x_n y_n},$$
$$\sqrt{x_1 + x_2 + \cdots + x_n} \times \sqrt{y_1 + y_2 + \cdots + y_n}.$$

   Show that for any sequence, $f \le g$.

11) Write a function which accepts a sequence $(a_1, a_2, \ldots, a_n)$ and returns $(a_1 + a_2 + \cdots + a_n)/n$. Use this function to demonstrate Chebycheff's Inequality, i.e.,

$$\frac{a_1 + a_2 + \cdots + a_n}{n} \times \frac{b_1 + b_2 + \cdots + b_n}{n} \le \frac{a_1 b_1 + a_2 b_2 + \cdots + a_n b_n}{n},$$

   for non-increasing sequences of numbers.

12) A number is called a *Harshad* number if it is divisible by the sum of its digits (e.g., 12 is Harshad as it is divisible by $1 + 2 = 3$). Find all 2-digit Harshad numbers. How many 5-digit Harshad numbers are there? (Harshad means "giving joy" in Sanskrit, defined and named by the Indian mathematician D. Kaprekar.)

13) The formula $e_{41} = n^2 + n + 41$ produces prime numbers for $0 \le n \le 39$ but not for $n = 40$. Check that for no i between 1 and 10000 does the formula $e_i = n^2 + n + i$ produce prime numbers for a larger interval starting at $n = 0$

14) Recall that Fibonacci numbers are defined recursively by $a_1 = a_2 = 1$ and $a_n = a_{n-1} + a_{n-2}$, for $n \ge 2$. Show that the following identities hold for $n = 100$.

$$a_1 + a_2 + \cdots + a_n = a_{n+2} - 1,$$
$$a_n^4 - a_{n-2} a_{n-1} a_{n+1} a_{n+2} = 1,$$
$$a_{2n} = a_n(a_{n-1} + a_{n+1}).$$

15) Compute

$$\sqrt{1 + \frac{1}{1^2} + \frac{1}{2^2}} + \sqrt{1 + \frac{1}{2^2} + \frac{1}{3^2}} + \cdots + \sqrt{1 + \frac{1}{2022^2} + \frac{1}{2023^2}}.$$

16) Investigate whether the series

$$2^{-\frac{1}{2}} + (3 + 5)^{-\frac{1}{2}} + (7 + 11 + 13)^{-\frac{1}{2}} + (17 + 19 + 23 + 29)^{-\frac{1}{2}} + \cdots$$

converges.

17) Write

$$\prod_{n=1}^{10} \frac{(2n - 1)(2n + 1)}{2n \times 2n} = \frac{1 \times 3}{2 \times 2} \frac{3 \times 5}{4 \times 4} \frac{5 \times 7}{6 \times 6} \cdots.$$

18) Investigate

$$\frac{e}{2} = \left(\frac{2}{1}\right)^{\frac{1}{2}} \left(\frac{2\ 4}{3\ 3}\right)^{\frac{1}{4}} \left(\frac{4\ 6\ 6\ 8}{5\ 5\ 7\ 7}\right)^{\frac{1}{8}} \left(\frac{8\ 10\ 10\ 12\ 12\ 14\ 14\ 16}{9\ 9\ 11\ 11\ 13\ 13\ 15\ 15}\right)^{\frac{1}{16}} \cdots$$

19) Investigate

$$\sum_{n=0}^{\infty} \frac{(-1)^n}{2n + 1} \sum_{k=0}^{2n} \frac{1}{2n + 4k + 3} = \frac{3\pi}{8} \log \frac{\sqrt{5} + 1}{2} - \frac{\pi}{16} \log 5.$$

20) Find the smallest number expressible as the sum of two cubes in two different ways (*hint:* the number is less than 3000).

Chapter 4
Functions

4.1 Functions

Functions in mathematics define rules about how to handle data. A function is a rule which assigns to each element in its domain a unique element in a specific range. For example, the function f defined as $f(n) = n^2 + 1$ will receive as an input (a number) n and its output will be $n^2 + 1$. Another way to think of this is that one can assign an object to the parameter n in the function and f will process the object according to the rules defined within the function.

Functions (in programming) provide a way to split the code into mini-programs with their own (local) variables and codes and one can pass inputs into them and receive outputs. In this way one can break a long program into logically smaller programs, each piece being a stand-alone function.

We start by defining simple functions in Python.

```
[1]: def f():
         print("Western Sydney University")
```

```
[2]: for i in range(4):
         f()
```

```
[2]: Western Sydney University
     Western Sydney University
     Western Sydney University
     Western Sydney University
```

Here f is the name of the function. What comes after : is the body of the function. One can think of f as a mini-program and one can call this function anytime in the main program. This function has a poor soul: no parameter to pass objects into it. Next we improve this function by passing data into it.

© The Author(s), under exclusive license to Springer Nature Switzerland AG 2023
R. Hazrat, *A Course in Python*, Springer Undergraduate Mathematics Series,
https://doi.org/10.1007/978-3-031-49780-3_4

```
[3]: def f(x):
         print(f'Western Sydney {x}')

     for i in ['School', 'University', 'Centre', 'Whatever!']:
         f(i)
```

```
[3]: Western Sydney School
     Western Sydney University
     Western Sydney Centre
     Western Sydney Whatever!
```

Here the function f comes with a parameter x. We can pass data to f via the parameter x. It is important to note that we have not specified any type for the objects passed into f via x. This opens up our hand.

Next we define a function name f accepting data (a variable) x and returning $x^2 + 1$, namely $f(x) = x^2 + 1$.

```
[4]: def f(x):
         return x**2 + 1
```

```
[5]: f(3)
```

```
[5]: 10
```

```
[6]: import math
     f(math.pi)
```

```
[6]: 10.869604401089358
```

```
[7]: f(f(2))
```

```
[7]: 26
```

```
[8]: f(3-2j) == (3-2j) * (3-2j) + 1
```

```
[8]: True
```

Note that we can pass to the function f data of type integers, floats or complex numbers, as the body of the function f is defined to do arithmetic on numbers. Later we see we can specify what type of data a function can accept.

Next we define the function $f(x) = \frac{1}{1+x}$ and then calculate

$$\frac{1}{1 + \frac{1}{1+x}},$$

for some values of x.

```
[9]:  def f(x):
          return 1 / (1 + x)
```

```
[10]:  f(1)
```

```
[10]:  0.5
```

One can see that $f(f(x)) = \frac{1}{1+\frac{1}{1+x}}$, thus we could evaluate the function with the composition

```
[11]:  f(f(1))
```

```
[11]:  0.6666666666666666
```

As one can see, the function is designed to accept (integer, float, complex) numbers. Python allows us to do symbolic computation, namely to work with a symbol x and carry out the arithmetic symbolically. For this we need to use the library sympy and specify that x is a symbol. Once this is done, Python can comfortably carry out the computations using this symbol.

```
[12]:  import sympy

       def f(x):
           return 1 / (1 + x)

       x=sympy.symbols('x')
```

```
[13]:  f(x)
```

$$[13]: \quad \frac{1}{x+1}$$

```
[14]:  f(f(f(x)))
```

$$[14]: \quad \frac{1}{1 + \frac{1}{1+\frac{1}{x+1}}}$$

```
[15]:  x=sympy.symbols('elephant')

       def g(x):
           return x /(1 + x)
```

```
[16]:  g(g(g(x)))
```

[16]:
$$\frac{elephant}{(elephant+1)\left(\frac{elephant}{elephant+1}+1\right)\left(\frac{elephant}{(elephant+1)\left(\frac{elephant}{elephant+1}+1\right)}+1\right)}$$

This makes it clear that in the definition of the function `f(x)`, the parameter `x` has no type assigned to it. So one can pass any object with any type into the function. It is the body of the function which determines what type one should pass into the function. As long as the functions are correctly designed for certain types, Python handles passing those types into the functions. The following example demonstrates this.

```
[17]: def f_image(x):
          display(x, x.convert('L'))
```

```
[18]: from PIL import Image
      y=Image.open('Napoleon.jpg')
```

```
[19]: f_image(y)
```

[19]:

Clearly this function cannot handle numbers:

Exercise 4.1 *Define* $f(x) = \sqrt{1+x}$ *in Python and show that*

$$f(f(f(f(f(x))))) = \sqrt{1 + \sqrt{1 + \sqrt{1 + \sqrt{1 + \sqrt{1+x}}}}}.$$

Solution

This exercise shows how comfortably Python can compose functions, something that becomes complicated and involved when done repeatedly.

```
[20]:  def f(x):
           return math.sqrt(1 + x)
```

```
[21]:  f(f(f(f(f(f(1))))))
```

```
[21]:  1.616121206508117
```

However, to approach this symbolically, we could employ the `sqrt` function in `sympy`:

```
[22]:  import sympy

       def sf(x):
           return sympy.sqrt(1 + x)
```

```
[23]:  x=sympy.symbols('x')

       sf(sf(sf(sf(sf(x)))))
```

$$[23]: \sqrt{\sqrt{\sqrt{\sqrt{\sqrt{x+1}+1}+1}+1}+1}$$

4.1.1 The scope of functions

Functions allow us to break a program into smaller, more manageable pieces. Each function governs its own *local* variables: the variables we define inside the functions. These local variables cannot be accessed from outside the function, so the scope of these variables is within the function they are defined. In contrast, the variables defined in the main body of the program are *global* variables, and these can be used throughout the code, also within the functions.

Exercise 4.2 *Define the functions* $p(n) = n(n+1)(n+2)(n+3) + 1$ *and* $q(n) = (n^2 + 3n + 1)^2$ *and observe that they are equal.*

Solution

Defining these functions is easy:

```
[24]: import math

      def p(n):
          #square_num is a local variable belonging to the function
          →p
          square_num = n * (n + 1) * (n + 2) * (n + 3) + 1
          return square_num

      def q(n):
          #square_num is a local variable belonging to the function
          →q
          square_num = (n**2 + 3 * n + 1) ** 2
          return square_num

      print(p(3), q(3))
```

```
[24]: 361 361
```

In this code, we have defined two functions p and q. Note that in both functions, there is a local variable square_num. These variables belong exclusively to their functions. One cannot call them from outside the function. Although they have the same name, they don't interfere with each other; one is defined and belongs to the function p and the other to q.

Exercise 4.3 *Consider the function*

$$ep(n) = 1 + \frac{1}{1} + \frac{1}{2!} + \cdots + \frac{1}{n!},$$

and observe that this series tend to Euler's number e.

Solution

This will be the first time that we define a variable within the body of a function. The function is called $ep(n)$ and inside it we define sum = 0. Note that this sum is a local variable, that is, it is only defined inside the function. We cannot call it from outside, which is a good thing.

```
[25]: import math

      def ep(n):
          s=0
          for i in range(n+1):
              s += 1/ math.factorial(i)
          return s
```

```
ep(10)
```

[25]: 2.7182818011463845

[26]:
```
for i in range(1, 100, 10):
    print(round(math.exp(1) - ep(i), 5), end =', ')
```

[26]: 0.71828, 0.0, -0.0, -0.0, -0.0, -0.0, -0.0, -0.0, -0.0, -0.0,

Exercise 4.4 *Write a function to calculate the following series*

$$f(n) = \frac{1}{1} + \frac{1}{1+2} + \ldots + \frac{1}{1+2+\ldots+n}.$$

Solution

We first write a function, called sumadd, which accepts n and returns $1+2+\cdots+n$.

[27]:
```
def sumadd(n):
    s = 0
    for i in range(n+1):
        s += i
    return s

sumadd(10)
```

[27]: 55

Now using sumadd we can translate the function $f(n)$ into Python.

[28]:
```
def ssum(n):
    ssum = 0
    for i in range(1, n+1):
        ssum += 1/sumadd(i)
    return ssum

print(ssum(3)," and ", 1 + 1/(1 + 2) + 1/(1 + 2 + 3))
```

[28]: 1.5   and   1.5

Exercise 4.5 *Plot the graphs of the functions* $f(x) = 2\exp(-x^2)$ *and* $g(x) = \cos(\sin(x) + \cos(x))$ *between* $[-\pi, \pi]$.

Solution

We first define the functions f and g and then compute them in the range $[-\pi, \pi]$. We will systematically work with the library `matplotlib` in Chapter 7. Here we will only use it to plot the graph. The commands used in the code related to the plot are rather self-explanatory and used to 'decorate' the output graph.

```
[29]:  from math import cos, sin, pi, exp
       import matplotlib.pyplot as plt

       def f(x):
           return 2 * exp(-x**2)
       def g(x):
           return cos(sin(x) + cos(x))

       x = []
       fl = []
       gl = []
       for i in range(0, 100):
           s = -pi + i * 2 * pi/100
           x.append(s)
           fl.append(f(s))
           gl.append(g(s))

       plt.figure(figsize=(3, 3));
       plt.plot(x,fl)
       plt.plot(x,gl);
```

[29]: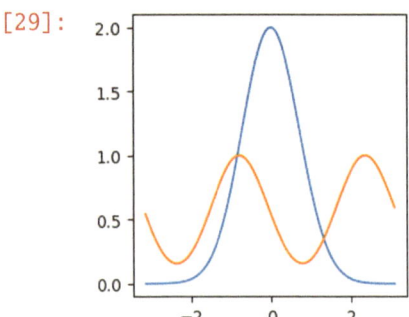

We can pass more than one object into functions. Here we write a function to check, for given n and m, whether $\sqrt{n^2 + m^2}$ is an integer, i.e., whether the pair (n, m) is a *Pythagorean pair*.

```
[30]:  import math

       def p(x,y):
           return math.sqrt(x**2 + y**2).is_integer()
```

```
[31]:  p(3, 4)
```

```
[31]:  True
```

```
[32]:  p(5, 3)
```

```
[32]:  False
```

Using the function p we can easily find the Pythagorean pairs, here up to 20.

```
[33]:  for n in range(1 , 21):
           for m in range(n, 21):
               if p(n, m):
                   print(n, m)
```

```
[33]:  3 4
       5 12
       6 8
       8 15
       9 12
       12 16
       15 20
```

We have seen p(x) accepts one object, whereas p(x, y) can receive two objects. If we need to pass a sequence of data into a function, one brilliant method is to do just that, passing a sequence into a function. The following example shows how this is done: We write the function

$$f(x_1, x_2, \ldots, x_n) = x_1^2 + x_2^2 + \cdots + x_n^2.$$

```
[34]:  def f(x, y):
           return x**2 + y**2
```

```
[35]:  f(2, 3)
```

```
[35]:  13
```

The input f(2, 3, 4) would generate an error.

```
[36]: def f(*x):
          print(x)
```

```
[37]: f(1, 2)
```

```
[37]: (1, 2)
```

```
[38]: f(1, 3, 5, 6)
```

```
[38]: (1, 3, 5, 6)
```

```
[39]: f(3, 'test', 1.4)
```

```
[39]: (3, 'test', 1.4)
```

With this approach, the function takes the form:

```
[40]: def f(*x):
          s = 0
          for i in x:
              s += i**2
          return s

      f(1, 2, 3)
```

```
[40]: 14
```

```
[41]: f(*range(4))
```

```
[41]: 14
```

Exercise 4.6 *Write the function* $f(x_1, x_2, \ldots, x_n) = \sqrt{x_1 + \sqrt{x_2 + \sqrt{x_3 + \cdots + \sqrt{x_n}}}}$ *and calculate* $f(1, 2, \ldots, 10)$.

Solution

We define a function which accepts the sequence $(x_1, x_2, \cdots, x_n)$ and then calculates the expression. A priori we don't know how long the sequence passing into the function is.

```
[42]: def f(*x):
          s = 0
          for i in x:
              s = math.sqrt(s) + i
```

```
    return math.sqrt(s)

f(1,2,3)
```

[42]: 2.1753277471610746

Going through the code, f(1,2,3) will assign the tuple $(1, 2, 3)$ to the parameter x. Next we run i through the tuple $(1, 2, 3)$. In the first run, $i = 1$ and we obtain $\sqrt{0} + 1$, which we assign to sum. In the next round of the loop, $i = 2$ and with the previous sum, we obtain $\sqrt{\sqrt{0} + 1} + 2$. Simplifying, this is $\sqrt{1} + 2$. Next, for $i = 3$, we get $\sqrt{\sqrt{1} + 2} + 3$. The loop is complete and the next line return math.sqrt(sum) will take another root for the result, i.e. we obtain $\sqrt{\sqrt{\sqrt{1} + 2} + 3}$. Rearranging, we obtain

$$\sqrt{3 + \sqrt{2 + \sqrt{1}}}.$$

Looking at the statement of the exercise, we wanted to get

$$\sqrt{1 + \sqrt{2 + \sqrt{3}}}.$$

Let us investigate this symbolically. We use sympy methods to define the function.

[43]:
```
def sf(*x):
    s = 0
    for i in x:
        s = sympy.sqrt(s) + i
    return sympy.sqrt(s)
```

[44]:
```
import sympy

x, y, z = sympy.symbols('x_1 x_2 x_3')
```

[45]: sf(x, y, z)

[45]:
$$\sqrt{x_3 + \sqrt{\sqrt{x_1} + x_2}}$$

But if we pass the reverse of this sequence into the function, i.e.:

[46]:
```
x, y, z = sympy.symbols('x_3 x_2 x_1')
```

[47]: sf(x, y, z)

[47]:

$$\sqrt{x_1 + \sqrt{x_2 + \sqrt{x_3}}}$$

then we get the right output. This clearly shows we have to start the loops from the right-hand side of sequence rather than the left-hand side. We have seen how to reverse lists or tuples.

[48]:
```
def sf(*x):
    s = 0
    for i in x[ : : -1]:
        s = sympy.sqrt(s) + i
    return sympy.sqrt(s)
```

[49]: `sf(x,y,z)`

[49]:

$$\sqrt{x_3 + \sqrt{\sqrt{x_1} + x_2}}$$

Exercise 4.7 *Given non-negative integers* $c_0, c_1, \ldots, c_m$, *with* $c_m \neq 0$, *define the function*

$$f(c_0, c_1, \ldots, c_m) = c_0 + \cfrac{1}{c_1 + \cfrac{1}{\cdots + \cfrac{1}{c_m}}}.$$

Solution

Here, similar to the previous exercise, we need to start with the last fraction, $\frac{1}{c_m}$, and work our way up to c_0. The next step in the process (a loop) would be $c_{m-1} + \frac{1}{c_m}$ and then $c_{m-2} + \frac{1}{c_{m-1}+\frac{1}{c_m}}$. This process can be captured by the code

```
s = 0
for i in c[ : : -1]:
    s = i + 1/s
```

As usual, in the first round of the loop, the sum s would start with 0. However, in that case we get an error message as we are using 1/s. A smart way to avoid this is to define an if statement inside the code to check if the value of s is zero, and in that case we ignore 1/s.

[50]:
```
def f(*c):
    s=0
    for i in c[ : : -1]:
        s = i + (1/s if s !=0 else 0)
    return s
```

```
[51]: f(1, 2, 3, 4)
```

```
[51]: 1.4333333333333333
```

```
[52]: f(*range(1,100))
```

```
[52]: 1.4331274267223117
```

Again, we can run the code symbolically to check if the process is working the way we had in mind.

```
[53]: import sympy

      x, y, z = sympy.symbols('x_1 x_2 x_3')
```

```
[54]: f(x, y, z)
```

$$[54]: \quad x_1 + \cfrac{1}{x_2 + \cfrac{1}{x_3}}$$

We need a list of symbols of the form $x_0, x_1, \ldots, x_m$ for a given m. The command `symbols('x:m')` where m is a positive integer does that.

```
[55]: s = sympy.symbols('x:11')
```

```
[56]: f(*s)
```

$$[56]: \quad x_0 + \cfrac{1}{x_1 + \cfrac{1}{x_2 + \cfrac{1}{x_3 + \cfrac{1}{x_4 + \cfrac{1}{x_5 + \cfrac{1}{x_6 + \cfrac{1}{x_7 + \cfrac{1}{x_8 + \cfrac{1}{x_9 + \frac{1}{x_{10}}}}}}}}}}$$

Here is yet another way to write this function.

```
[57]: def f(L):
          L.reverse()
          s = L[0]
          for i in range(1,len(L)):
              s = L[i] + 1/s
          return s
```

```
[58]: f([1, 2, 3, 4])
```

```
[58]: 1.4333333333333333
```

We will revisit this exercise and write a code using a functional programming approach.

Exercise 4.8 *Define a sequence of numbers by $a_0 = 1$, $a_1 = 1$ and, for $n \geq 2$, $a_n = 3a_{n-1} - a_{n-2}$. Write a function to accept n and calculate a_n, for $0 \leq n \leq 10$.*

Solution

The function a_n or rather, $a(n)$, is a recursive function, that is, it calls itself. We write two different codes for this. The first one is a direct translation of $a(n)$ into Python.

```
[59]: def a(n):
          if n == 0 or n ==1:
              return 1
          else:
              return 3 * a(n - 1) - a(n - 2)

      for n in range(11):
          print(a(n), end=' ')
```

```
[59]: 1 1 2 5 13 34 89 233 610 1597 4181
```

Next we give an alternative code. Here the code does not immediately look like a recursive function.

```
[60]: def a(n):
          x, y =1, 1
          for i in range(n-1):
              x, y =y, 3*y - x
          return y

      for n in range(11):
          print(a(n), end=' ')
```

```
[60]: 1 1 2 5 13 34 89 233 610 1597 4181
```

Exercise 4.9 *Write a function to check if a number is prime.*

Solution

The function isprime from the sympy library is designed to check if a number is prime.

```
[61]:  from sympy import isprime

       print(f'Is 17*23 a prime number? {isprime(17*23)}')
```

```
[61]:  Is 17*23 a prime number? False
```

Here we write our own function to determine the primeness of a positive integer.

```
[62]:  def primeQ(n):
           if n == 0 or n == 1 or type(n) != int:
               return False
           for i in range(2, n//2 + 1):
               if n % i == 0:
                   return False
           return True
```

```
[63]:  primeQ(2)
```

```
[63]:  True
```

```
[64]:  primeQ(13*17)
```

```
[64]:  False
```

Next we check whether the Mersenne number $2^{31} - 1$ is prime. We do this both using our function and the built in function isprime from the sympy library. We can also time the whole operation to see how long it actually takes to complete the job. This can be done by importing the function time from a library with the same name.

```
[65]:  import time

       start_time = time.time()
       print(primeQ(2**31 - 1))
       end_time = time.time()
       print("%s seconds" %(end_time - start_time))
```

```
[65]:  True
       59.52177691459656 seconds
```

```
[66]:  start_time = time.time()
       print(isprime(2**31 - 1))
       end_time = time.time()
       print("%s seconds" %(end_time - start_time))
```

[66]: True
0.0055501461029052734 seconds

This little experiment shows there is a lot of room to improve our code for the function primeQ!

Exercise 4.10   1. *Write a function to accept a sequence of numbers and then print the sequence which only contains the primes of the original sequence.*

2. *Write a function to accept a sequence of numbers and then print the sequence up to the occurrence of the first prime.*

Solution

Python offers a technique called list comprehension. Using list comprehension, one can very easily and elegantly write the above codes. Here, however, we write a code with the techniques we have studied so far.

```
[67]: def all_primes(l):
          L = []
          for i in l:
              if primeQ(i):
                  L.extend([i])
          return L
```

```
[68]: all_primes([1, 4, 199, 2, 5, 7, 11, 2, 4, 3])
```

```
[68]: [199, 2, 5, 7, 11, 2, 3]
```

For the next part of the Exercise:

```
[69]: def up_prime(l):
          i = 0
          while not primeQ(l[i]):
              i += 1
          return l[ : i]
```

```
[70]: up_prime([1, 10, 8, 8, 7, 2, 5, 7, 11, 2, 4, 3])
```

```
[70]: [1, 10, 8, 8]
```

```
[71]: up_prime([5, 10, 8, 8, 7, 2, 5, 7, 11, 2, 4, 3])
```

```
[71]: []
```

Here is another approach, using the sequence to pass the list inside the function and then using the **break** once the first prime number has been spotted..

```
[72]: def up_prime2(*n):
          for i in range(len(n)):
              if primeQ(n[i]):
                  break
          return n[ : i]
```

```
[73]: up_prime2(1, 10, 8, 8, 7, 2, 5, 7, 11, 2, 4, 3)
```

```
[73]: (1, 10, 8, 8)
```

```
[74]: up_prime2(5, 10, 8, 8, 7, 2, 5, 7, 11, 2, 4, 3)
```

```
[74]: ()
```

4.1.2 Functions, default values

We have seen that one can pass objects into functions. We now design functions for which the objects are pre-defined. This means, if one does not pass any object for a given parameter, the pre-defined values will be used.

```
[75]: def address(name, career, city):
          return print(f' {name} is a {career} who lives in␣
      ␣{city}')
```

```
[76]: address('Whitlam', 'politician', 'Sydney')
```

```
[76]:  Whitlam is a politician who lives in Sydney
```

Next we modify the function address, giving default values to the parameters, which will be used if those parameters are not assigned any object by the user.

```
[77]: def address(name, career = 'politician', city = 'Sydney'):
          return print(f'{name} is a {career} who lives in {city}')
```

```
[78]: address('Gladys Brejeklian')
```

```
[78]: Gladys Brejeklian is a politician who lives in Sydney
```

```
[79]: address('Daniel Andrews', city = 'Melbourne')
```

```
[79]: Daniel Andrews is a politician who lives in Melbourne
```

```
[80]: address('Anthony Albanese', career = 'prime minster', city =␣
       ↪'Capital territory')
```

```
[80]: Anthony Albanese is a prime minster who lives in Capital
      territory
```

Exercise 4.11 *Write a function to ask for the values of a, b and n, and produce the n-th Fibonacci number with the initial values a and b, i.e., $f_0 = a$, $f_1 = b$ and $f_n = f_{n-1} + f_{n-2}$. Further, modify the function so that if the user does not enter the initial values a or b, then they take the default values $a = 1$ and $b = 1$.*

Solution

First we write the code of the function without specifying any initial values. The approach in generating the Fibonacci sequence in this code is similar to Exercise 4.8.

```
[81]: def fibonacci(N, a, b):
          for i in range(N-2):
              a, b = b, a + b
          return b
```

```
[82]: fibonacci(10, 1, 1)
```

```
[82]: 55
```

```
[83]: def fibonacci(N, a = 1, b = 1):
          for i in range(N-2):
              a, b = b, a + b
          return b
```

```
[84]: fibonacci(10)
```

```
[84]: 55
```

```
[85]: fibonacci(3, b = 15)
```

```
[85]: 16
```

4.1.3 Functions, specific types

As we discussed, the function def f(x): can accept any type of object. If we would like to restrict the type of object that f can handle from the outset, we can specify the type of x.

```
[86]:  def f(x : list, i : int):
           if i <= len(x):
               return x[ : i]
```

```
[87]:  f(['pick', 'the', 'first', 'i-th', 'elements', 'of', 'the',␣
       ↪'list'], 5)
```

```
[87]:  ['pick', 'the', 'first', 'i-th', 'elements']
```

Here the function f *expects* the parameter to be passed into it to be of the type list and integer. This way of writing code also helps us to understand what a code does when first reading it.

```
[88]:  f('hello here he comes', 4)
```

```
[88]:  'hell'
```

4.2 Functional Programming: Anonymous (lambda) Functions

Sometimes we need to 'define a function as we go' and use it on the spot. Python enables us to define a function without giving it a name, use it, and then move on! These functions are called *anonymous* functions. Obviously if we need to use a specific function frequently, then the best approach is to give it a name and define it, as we did earlier. Here is an anonymous function equivalent to $f(x) = x^2 + 4$:

```
[89]:  lambda x : x**2 + 4
```

```
[89]:  <function __main__.<lambda>(x)>
```

```
[90]:  (lambda x : x**2 + 4)(10)
```

```
[90]:  104
```

With the keyword lambda we define an anonymous function. Here the function has one variable, x, and its output is $x^2 + 4$. One can think of this as $x \mapsto x^2 + 4$, where the lambda is that arrow.

The following might defeat the purpose: giving a name to an anonymous function. But let's make sure we are comfortable with the concept.

```
[91]:  f = lambda x: x + 3
```

```
[92]:  f(1)
```

[92]: 4

One can define anonymous functions with several variables.

[93]: ```
(lambda x, y: x + y)(2, 3)
```

[93]:  5

[94]:  ```
(lambda x, y, z: x + y + z)(1, 2, 3)
```

[94]: 6

[95]: ```
(lambda x, y, z: x + y + z)('one', ' two', ' three')
```

[95]:  'one two three'

As the above two examples show, as long as the body of the function can handle the types, one can pass any type of object into the parameters of the function.

Similar to the 'classical' functions, we can define default values for parameters.

[96]:  ```
(lambda x, y, z = 3: x * y * z)(1, 2)
```

[96]: 6

[97]: ```
(lambda x, y = ' prepared', z = ' pizza': x + y + z)('You')
```

[97]:  'You prepared pizza'

[98]:  ```
(lambda x, y = ' prepared', z = ' pizza': x + y + z)('I', z =
    ' hamburger')
```

[98]: 'I prepared hamburger'

[99]: ```
(lambda x, y = ' prepared', z = ' pizza': x + y + z)('He', y
    = ' burnt')
```

[99]:  'He burnt pizza'

Recall that via * we can pass sequences of objects into functions.

[100]:  ```
(lambda *z: z[ : : -1])(1, 2, 3)
```

[100]: (3, 2, 1)

[101]: ```
(lambda *z: z[: : -1])(*range(13))
```
```

```
[101]:  (12, 11, 10, 9, 8, 7, 6, 5, 4, 3, 2, 1, 0)
```

```
[102]:  (lambda *z: z + ('again',) + z)(1, 2, 3)
```

```
[102]:  (1, 2, 3, 'again', 1, 2, 3)
```

```
[103]:  sum([1,2,3])
```

```
[103]:  6
```

```
[104]:  (lambda *args: sum(args))(1,2,3)
```

```
[104]:  6
```

In Python there are many ways to construct a collection of objects, for example via list, tuple, dictionary or iterators. There are times when we would like to apply a function to all the objects of a list or a collection. Suppose f is a function and [a,b,c] is a list. We want to be able to *push* the function f inside the list and get [f(a),f(b),f(c)]. This can be done using the command map.

As a first example, we map the function sin from the math library to the list of integers 1 to 9.

```
[105]:  from math import sin, cos, pi

        x = map(sin, range(1, 10))
        xlist = list(x)
        print(xlist)
```

```
[105]:  [0.8414709848078965, 0.9092974268256817, 0.1411200080598672,
         -0.7568024953079282, -0.9589242746631385,
         -0.27941549819892586, 0.6569865987187891, 0.9893582466233818,
         0.4121184852417566]
```

We can immediately write amusing codes!

```
[106]:  import matplotlib.pyplot as plt

        x = map(sin, range(1, 50))
        xlist = list(x)
        y = map(cos, range(1, 50))
        ylist = list(y)
        plt.axis('equal')
        plt.plot(xlist, ylist);
```

[106]: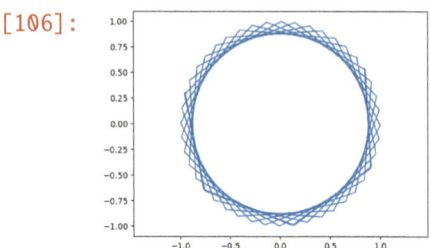

Oftentimes, one defines a function via lambda and then applies it to the list, as the example below demonstrates.

[107]:
```python
l = map(lambda x: sin(x * pi/2), range(10))
```

[108]:
```python
list(l)
```

[108]:
```
[0.0,
 1.0,
 1.2246467991473532e-16,
 -1.0,
 -2.4492935982947064e-16,
 1.0,
 3.6739403974420594e-16,
 -1.0,
 -4.898587196589413e-16,
 1.0]
```

Of course we could create this list via a for loop, but there are times when functional programming is much shorter, more elegant and easier to read.

[109]:
```python
L=[]
for x in range(10):
    L.append(sin(x * pi/2))
L
```

[109]:
```
[0.0,
 1.0,
 1.2246467991473532e-16,
 -1.0,
 -2.4492935982947064e-16,
 1.0,
 3.6739403974420594e-16,
 -1.0,
 -4.898587196589413e-16,
 1.0]
```

Here are some more examples showing the versatility of this approach.

```
[110]: from sympy import symbols, factor, expand

       x=symbols('x')
       y=symbols('y')
```

We create the expansion of $(x + y)^i$, for $1 \leq i \leq 5$.

```
[111]: lx = map(lambda i: expand((x + y)**i), range(1,5))
```

```
[112]: list(lx)
```

```
[112]: [x + y,
        x**2 + 2*x*y + y**2,
        x**3 + 3*x**2*y + 3*x*y**2 + y**3,
        x**4 + 4*x**3*y + 6*x**2*y**2 + 4*x*y**3 + y**4]
```

Next we use the map function to rotate a picture certain number of times.

```
[113]: from PIL import Image
       x=Image.open('Napoleon.jpg')
```

```
[114]: x_pic=map(lambda n: display(x.rotate(n * 20)), range(1, 5));
```

```
[115]: list(x_pic);
```

[115]:

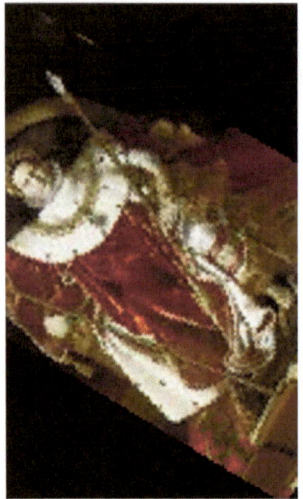 

Exercise 4.12 *Recall that the formula* $n(n + 1)(n + 2)(n + 3) + 1$ *produces a square number. Using functional programming, calculate this number for* $1 \leq n \leq 10$ *and its square root.*

Solution

Let us define a function $f(n)$ which gives back $(n, \quad n(n + 1)(n + 2)(n + 3) + 1)$. We then map this function inside the list of integers from 1 to 10.

```
[116]:  import math

        def f(n):
            x=n * (n + 1) * (n + 2) * (n + 3) + 1
            return x , math.sqrt(x)

        list(map(f, range(1,11)))
```

```
[116]:  [(1, 1.0),
         (25, 5.0),
         (121, 11.0),
         (361, 19.0),
         (841, 29.0),
         (1681, 41.0),
         (3025, 55.0),
         (5041, 71.0),
         (7921, 89.0),
         (11881, 109.0),
         (17161, 131.0)]
```

Here is another example, showing practically anything can be seen as a function and thus sent into a list via map. We will be using the plotting facility of the sympy library here. The function pl.plot(f) will plot the function f. We define an anonymous function lambda f : pl.plot(f) and then we replace f with different functions in a list, via mapping the lambda function inside the list. This allows us to plot all the graphs within one line of code!

[117]:
```
from sympy import sin, cos, symbols
from sympy import plotting as pl

x = symbols('x')
l = map(lambda f : pl.plot(f), (sin(x), x + sin(x)**3, cos(x)⎵
  ⤶+ sin(x)**3))
list(l);
```

[117]:

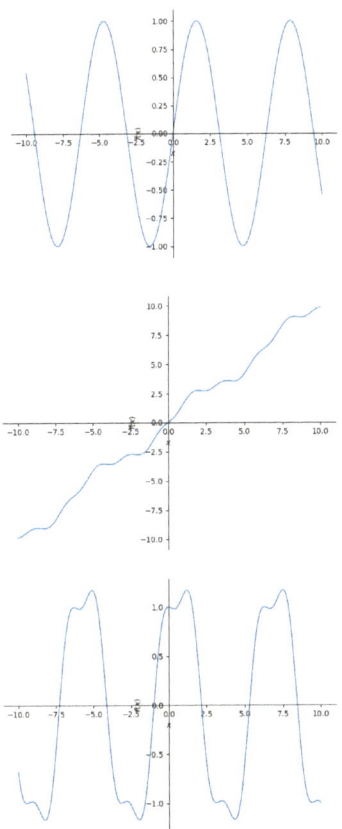

We will discuss in depth the concepts of iterators and generators in Python in Chapter 5. Here we only mention that, when we use map to apply a function, say, f, to a list or any ordered set, Python actually does not immediately apply f to all the elements of the set. Each time we prompt the map statement, it will apply the function f to the set in order. Of course by invoking list we can obtain the entire action, as we have done so far (see Problem 14).

4.2.1 Selecting from a collection

So far we have been able to create a collection of objects, and apply a function to each of the objects. The next step is to be able to choose, from a collection of objects, certain objects which fit a specific description. This can be achieved by using the command filter, as the following example demonstrates.

How many numbers of the form $3n^5 + 11$, when n varies from 1 to 500, are prime?

We define the anonymous function lambda n : 3 * n**5 + 11 and map it to the list of the first 500 positive integers.

```
[118]: s = map(lambda n : 3 * n**5 + 11, range(1, 501))
```

Next we use the function isprime from the sympy library to check which of these numbers are prime. To select, among these lists, those which are prime, we use filter.

```
[119]: from sympy import isprime

p = filter(isprime, s)
```

```
[120]: x = list(p)
```

```
[121]: print(x, sep=' , ')
```

```
[23]: [107, 3083, 23339, 746507, 1613483, 3145739, 72900011,
136306283, 617888939, 1140612107, 7606576139, 14112810539,
36499587083, 52870250507, 63010249739, 87948751883,
95273908139, 243410436107, 451609936907, 824384664683,
867763964939, 1691848015883, 2594599836683, 5350160323307,
8048749055339, 13630627200011, 18139852800011,
40149811396907, 42089062300139, 52941415424939,
73308784432139, 86486190621707]
```

```
[122]: len(x)
```

```
[122]: 32
```

In a nutshell, filter(isprime, s) will apply the function isprime (which returns True or False) to all the elements of s, and then collects those elements for which

isprime(x) is true.
In mathematical syntax, this is

$$\{x \in s \mid isprime(x)\}.$$

Exercise 4.13 *For which* $1 \leq n \leq 1000$ *does the Mersenne formula* $2^n - 1$ *produce a prime number.*

Solution

Once we know how to combine lambda function with filter this code is very easy to put together.

```
[123]: s = filter(lambda n : isprime(2**n - 1), range(1, 1001))
```

```
[124]: list(s)
```

```
[124]: [2, 3, 5, 7, 13, 17, 19, 31, 61, 89, 107, 127, 521, 607]
```

Let us take a deep breath and go through this one-liner slowly. The function lambda
n : isprime(2**n - 1) is an anonymous function which returns True if the
number $2^n - 1$ is prime and False otherwise. That is, the output of this anonymous
function is a boolean value. We thus can use filter with this function.

```
[125]: (lambda n : isprime(2**n - 1))(13)
```

```
[125]: True
```

range(1,1001) creates a list containing the numbers from 1 to 1000. The
command filter applies the anonymous function lambda n : isprime(2**n
- 1) to each element of this list and when the result is true, i.e.,
when $2^n - 1$ is prime, the element n will be selected. The numbers
$\{2, 3, 5, 7, 13, 17, 19, 31, 61, 89, 107, 127, 521, 607\}$ are the only n's in the given
range for which $2^n - 1$ is a prime number.

Just to give a comparison of functional programming with standard procedural
programming, we can write the following code which does the same thing.

```
[126]: L=[]
       for i in range(1, 1001):
           if isprime(2**i - 1):
               L.append(i)
       L
```

```
[126]: [2, 3, 5, 7, 13, 17, 19, 31, 61, 89, 107, 127, 521, 607]
```

Exercise 4.14 *Find the number of positive integers $0 < n < 20000$ such that 1997 divides $n^2 + (n + 1)^2$. Try the same code for 2023.*

Solution

We do this in one go.

```
[127]:  len(list(filter(lambda n: (n**2+ (n+1)**2) % 1997 == 0,␣
          ↪range(1,20000)))))
```

[127]: 20

First things first, we define a lambda function which checks, for a number n, whether the expression $n^2 + (n + 1)^2$ is divisible by 1997. This can be done as follows:

```
[128]:  lambda n: (n**2+ (n+1)**2) % 1997 == 0
```

[128]: <function __main__.<lambda>(n)>

The above function is a Boolean function and will return `True` or `False`.

```
[129]:  (lambda n: (n**2+ (n+1)**2) % 1997 == 0)(13)
```

[129]: False

This means $13^2 + (13 + 1)^2$ is not divisible by 1997. Next we use this function, with `filter`, and pass it into the `range(1,20000)`. We remind the reader that

```
filter(lambda n: (n**2+ (n+1)**2) % 1997 == 0, range(1,20000))
```

is an iterator and in itself will not perform the calculations. It needs us to prompt it to start the calculations. The command `list` will do just that, and the iterator will go through the whole range and create a list. The command `len` then gives us the length of this list.

Exercise 4.15 *Using functional programming, calculate $1^3 + 2^3 + \cdots + 13^3$.*

Solution

To add the numbers together we can use the function `sum`.

```
[130]:  sum([1,2,3])
```

[130]: 6

```
[131]:  l = map(lambda n : n**3, range(1, 14))
```

```
[132]: list(l)
```

```
[132]: [1, 8, 27, 64, 125, 216, 343, 512, 729, 1000, 1331, 1728,
        30664297]
```

```
[133]: sum(list(map(lambda n : n**3, range(1, 14))))
```

```
[133]: 30670381
```

To compare with the standard codes one could write:

```
[134]: s = 0
       for i in range(1,14):
           s += i**3
       print(s)
```

```
[134]: 30670381
```

4.2.2 Functional programming, reduce

Finally we introduce the reduce function. Recognizing how to use this function allows us to create very elegant and short programs. The function reduce works as follows:

$$reduce(f,[x,y,z,t]) \; gives \; f(f(f(x, \; y), \; z), \; t)$$

Here f is a function and [x,y,z,t] is a list of objects. The function reduce is in the library functools, so one needs to import it before using it.

```
[135]: from functools import reduce
```

As a first demonstration, given $\{x_1, x_2, \cdots, x_n\}$ we use reduce to produce $\{x_1 + x_2 + \cdots + x_n\}$.

```
[136]: reduce(lambda x, y: x + y, [1, 2, 3, 4, 5])
```

```
[136]: 15
```

If we follow the code, we see the program first assigns 1 and 2 to x and y respectively, and returns $x + y$, which is $1 + 2$. Next this will be assigned to x and y takes the next item in the list, namely 3 and again $x + y$ returns $1 + 2 + 3$. Repeating this, we get the sum of all the digits in the list.

```
[137]:  reduce(lambda x, y: x + ' ' + y,↵
        ↳["Western","Sydney","University"])
```

```
[137]:  'Western Sydney University'
```

```
[138]:  from sympy import symbols
```

```
[139]:  x, y =symbols('x y')
```

```
[140]:  reduce(lambda a, b: a ** b, [x, y, x, y, x])
```

$$[140]:\ \left(\left(\left(x^{y}\right)^{x}\right)^{y}\right)^{x}$$

```
[141]:  reduce(lambda a, b: a + b, symbols('x:10'))
```

$$[141]:\ x_0 + x_1 + x_2 + x_3 + x_4 + x_5 + x_6 + x_7 + x_8 + x_9$$

```
[142]:  reduce(lambda a, b: a * b, symbols('x:10'))
```

$$[142]:\ x_0 x_1 x_2 x_3 x_4 x_5 x_6 x_7 x_8 x_9$$

We revisit Exercise 4.7 and employ reduce to write the code.

Exercise 4.16 *Given non-negative integers $c_0, c_1, \ldots, c_m$, where $c_0 \neq 0$, define the function*

$$f(c_0, c_1, \ldots, c_m) = c_m + \cfrac{1}{c_{m-1} + \cfrac{1}{\cdots + \cfrac{1}{c_0}}}.$$

Solution

The code is short, but it requires a bit of concentration to see how it works. The best approach is to work out the first couple of steps by hand to observe that the result takes the form of the continued fraction described in the exercise.

```
[143]:  from functools import reduce

        L=[1, 2, 3, 4]
        L.reverse()

        reduce(lambda x, y: (1/x) + y, L)
```

```
[143]:  1.4333333333333333
```

Here we produce this continued fraction using sympy. We need a list of symbols of the form $c_0, c_1, \ldots, c_m$ for a given m. The command symbols('c:m'), where m is a positive integer, does that.

```
[144]: symbols('c:10')
```

```
[144]: (c0, c1, c2, c3, c4, c5, c6, c7, c8, c9)
```

We put this and the code above together:

```
[145]: from functools import reduce
       from sympy import symbols

       def L(m):
           sym = 'c:' + str(m)
           return symbols(sym)

       reduce(lambda x, y: (1/x) + y, L(10))
```

[145]:
$$c_9 + \cfrac{1}{c_8 + \cfrac{1}{c_7 + \cfrac{1}{c_6 + \cfrac{1}{c_5 + \cfrac{1}{c_4 + \cfrac{1}{c_3 + \cfrac{1}{c_2 + \cfrac{1}{c_1 + \frac{1}{c_0}}}}}}}}}$$

Exercise 4.17 *Find all the numbers up to one million which have the following property: if $n = d_1 d_2 \cdots d_k$ then $n = d_1! + d_2! + \cdots + d_k!$ (e.g. $145 = 1! + 4! + 5!$).*

Solution

We first need to get the digits of a given number. One way to do this is to convert the number into a string, get the list of all the letters, and then convert all the letters back to numbers.

```
[146]: str(1234)
```

```
[146]: '1234'
```

```
[147]: list(str(1234))
```

```
[147]: ['1', '2', '3', '4']
```

```
[148]: dig = map(int, list(str(1234)))
```

Now that we have all the digits, using map we can push the factorial function inside the list and calculate $n!$ for each digit.

```
[149]:  import math

        dig2 = map(lambda n: math.factorial(n), dig)
```

Finally, we add all the results together and, as we have seen, we can use the elegant reduce to do so.

```
[150]:  reduce(lambda x, y: x + y, dig2)
```

[150]: 33

Putting all these together, we can write a function.

```
[151]:  def f(n):
            dig = map(int, list(str(n)))
            dig2 = map(lambda n: math.factorial(n), dig)
            return reduce(lambda x, y: x + y, dig2)
```

```
[152]:  f(1234)
```

[152]: 33

We are ready to use the function f to found out those numbers $n = d_1 d_2 \cdots d_k$ such that $n = d_1! + d_2! + \cdots + d_k!$

```
[153]:  for i in range(1,1000001):
            if i == f(i): print(i)
```

[153]: 1
        2
        145
        40585

Exercise 4.18 *Notice that* $12^2 = 144$ *and* $21^2 = 441$, *i.e., the numbers and their squares are reverses of each other. Find all the numbers up to* 10000 *with this property.*

Solution

We first define a function which reverse the digits of a given number. We use strings and benefit from the functions available for this object.

```
[154]:  def re(n):
            return int(str(n)[ : : -1])
```

```
[155]:  re(12345)
```

```
[155]:  54321
```

Having this function under our belt, the solution to the problem is just one line. Notice that the problem is asking for the numbers n such that $re[n**2]=re[n]**2$.

```
[156]:  a=list(filter(lambda n: re(n**2) == re(n)**2,
        ↪range(1,10001)))
```

```
[157]:  print(*a, sep = ', ')
```

```
[157]:  1, 2, 3, 10, 11, 12, 13, 20, 21, 22, 30, 31, 100, 101, 102,
        103, 110, 111, 112, 113, 120, 121, 122, 130, 200, 201, 202,
        210, 211, 212, 220, 221, 300, 301, 310, 311, 1000, 1001,
        1002, 1003, 1010, 1011, 1012, 1013, 1020, 1021, 1022, 1030,
        1031, 1100, 1101, 1102, 1103, 1110, 1111, 1112, 1113, 1120,
        1121, 1122, 1130, 1200, 1201, 1202, 1210, 1211, 1212, 1220,
        1300, 1301, 2000, 2001, 2002, 2010, 2011, 2012, 2020, 2021,
        2022, 2100, 2101, 2102, 2110, 2111, 2120, 2121, 2200, 2201,
        2202, 2210, 2211, 3000, 3001, 3010, 3011, 3100, 3101, 3110,
        3111
```

In the last exercise we will experiment with the amusing topic of finding secret messages in ancient texts and scriptures. Consider a text in the form of a string of letters $l_1 l_2 \cdots l_k$. Then an *equidistant letter sequence* of length s is a subsequence $l_n l_{n+d} \cdots l_{n+(s-1)d}$. Here d is called the *skip*. Note that this means taking letters from the text with uniform spacing. The subsequence can be thought of as a vertical section of text which appears when writing the text around a cylinder with a fixed circumference. The topic of finding hidden messages in scriptures by choosing the correct spacing has been a theme of many articles. Here we will look at

Exercise 4.19 *Search Jane Austen's* Sense and Sensibility, *for the word "google", which might appear as a equidistant letter sequence.*

Solution

Recall from 2.6.1 that the library nltk contains several classical texts as well as tools to process texts.

```
158]:  import nltk
       from nltk.book import *
```

```
158]:  *** Introductory Examples for the NLTK Book ***
       Loading text1, ..., text9 and sent1, ..., sent9
       Type the name of the text or sentence to view it.
       Type: 'texts()' or 'sents()' to list the materials.
       text1: Moby Dick by Herman Melville 1851
```

```
text2: Sense and Sensibility by Jane Austen 1811
text3: The Book of Genesis
text4: Inaugural Address Corpus
text5: Chat Corpus
text6: Monty Python and the Holy Grail
text7: Wall Street Journal
text8: Personals Corpus
text9: The Man Who Was Thursday by G . K . Chesterton 1908
```

[159]:
```
len(text2)
```

[159]: 141576

When we look inside the books, they are lists of strings (words). For our purpose we need to put add the characters together.

[160]:
```
text2[11 : 20]
```

[160]: ['The', 'family', 'of', 'Dashwood', 'had', 'long', 'been',↵
 ↪'settled', 'in']

This can be achieved by using join.

[161]:
```
s1 = 'Hello'
s2 = 'World'
```

[162]:
```
''.join([s1, s2])
```

[162]: 'HelloWorld'

Next we use reduce to get the strings, two at a time, and put them together.

[163]:
```
from functools import reduce

reduce(lambda x, y : ''.join([x,y]), text3[ :10])
```

[163]: 'InthebeginningGodcreatedtheheavenandtheearth'

[164]:
```
x_text = reduce(lambda x, y : ''.join([x,y]), text2)
```

Now we are ready to look at equidistant letter sequences. We can find the word 'google' in Jane Austen's book, published in 1811.

[165]:
```
word = 'google'
l_w = len(word)
for interval in range(1, 100):
```

```
        for j in range(len(x_text)):
            if x_text[0 + j : l_w * interval + j : interval].
↪lower() == 'google':
                print(f'{word} appears in the book starting from↳
↪the {j}th letter with interval {interval}')
```

[165]: google appears in the book starting from the 312852th letter
       with interval 90

[166]: x_text[312852: 6 * 90 + 312852 : 90]

[166]: 'googlE'

Problems

1. For integers $2 \leq n \leq 200$, find all n such that n divides $(n-1)! + 1$. Show that there are 46 such n.

2. Write a function which returns an answer of True if the n'th prime number is of the form $4k + 1$ for some integer k. Hence produce a list of those positive integers n lying between 1 and 100 such that the n'th prime is of such a form $4k + 1$. Verify that there are 47 such n.

3. Suppose P_i is the i-th prime number. Let $E_i = P_1 \times P_2 \times \cdots \times P_i + 1$ be the i-th Euclid number. Define a function Euclidnumber(n) to produce the n-th Euclid number. Use this function and produce the list of the first 20 Euclid numbers. Find out which of these numbers are in fact prime. Find the indices of the prime Euclid numbers (i.e, i, where E_i is prime), where $1 \leq i \leq 100$.

4. Write a function to calculate the following

$$p(n) = 1 + \frac{1}{\sqrt{1}} + \frac{1}{\sqrt{1} + \sqrt{2}} + \frac{1}{\sqrt{2} + \sqrt{3}} + \frac{1}{\sqrt{3} + \sqrt{4}} + \cdots + \frac{1}{\sqrt{n-1} + \sqrt{n}}$$

    (How many ways can you write this function?)

5. Let us call a number $a_0a_1...a_{n-1}a_n$ *pure prime* if $a_0a_1...a_{n-1}a_n$, $a_0a_1...a_{n-1} \cdots$, a_0a_1 and a_0 are all prime numbers. Find all the 3, 4 and 5 digit numbers which are pure prime. Can you conjecture how many pure prime 16 digit numbers exists?(Example: 7193 is a pure prime number, as 7193, 719, 71 and 7 are prime). Write the code using functional programming.

6. A function f of a natural number n is defined as follows. Let p_n denote the n'th prime number. Initialise s to be 1 and then replace it by $(s \times p_k) - 1$ for $k = 1, 2, \ldots, p_n$ in turn. The final value of s is the value of $f(n)$. The first few

values are $f(1) = 1$, $f(2) = 2$, $f(3) = 9$ and $f(4) = 62$. Define this function in Python. What are $f(5)$ and $f(10)$?

7. Let $A(x, y) = (a_{ij}(x, y))$ be the $n \times n$ matrix with

$$a_{ij}(x, y) = \begin{cases} 0 & \text{if } i = j \\ \sin(x)^i \cos(y)^j \left(\sin(y) + \cos(x) \right)^{i+j} & \text{if } i \neq j. \end{cases}$$

Create the 3×3 matrix $A(\pi, \pi/5)$.

8. The *norm* of a vector $x = (x_1, x_2, \ldots, x_n)$ of real numbers (i.e., floats) is defined to be

$$\|x\| = \sqrt{\sum_{k=1}^{n} x_k^2}.$$

Write a function vnorm which computes the norm of a real vector of any length. (Write the function in two different ways: using procedural loops and also using functional programming).

9. For a number n, a proper divisor k is a number which is not n and which divides n. For example, $\{1, 3, 5\}$ are all the proper divisors of 15. Consider the sum of all the proper divisors of a number. Now consider the sum of all the proper divisors of this new number and repeat the process. If one eventually obtains the number which one started with, then this number is called a *social number*. Write a program to show that 1264460 is a social number. Check whether 14316 is also a social number.

10. Using functional programming, demonstrate that for any sequence of positive numbers $a_1, a_2, \ldots, a_n$ we have

$$\left(a_1 + a_2 + \cdots + a_n\right)\left(\frac{1}{a_1} + \frac{1}{a_2} + \cdots + \frac{1}{a_n}\right) \geq n^2$$

11. We call a pair of prime numbers p and q *friends* if pq and qp are both prime (by pq we mean the juxtaposition of the numbers together). For example, the prime numbers 563 and 587 are friends as 563587 and 587563 are both primes. Write a program to produce all the prime friends for some suitable range of primes.

12. The number π starts with

$$3.14159265358979323846264338327950288419716939937510582097494$$
$$45923078164062862089986280348253421170679\cdots$$

and continues with no pattern. Search the first $100,000$ digits of π and check if your birthday appears as a sequence of digits in π. For exam-

ple the date Thursday 8th of April of 1971 appears in the digits of π:
3.14159265358979323846264338327950288$41971$693 $\cdots$

13. Recall that if $f : A \rightarrow B$ and $g : B \rightarrow C$ are two functions, then $gf : A \rightarrow C$
is defined by $gf(a) = g(f(a))$, where $a \in A$. Explain what the following code
does and rewrite it by using only one map.

```
[1]: z = map(lambda i : i+1, map(lambda i: i**2, range(1,10)))
```

```
[2]: list(z)
```

```
[3]: [2, 5, 10, 17, 26, 37, 50, 65, 82]
```

14. Recall that map creates an iterator that will be executed once it is prompted. Run
the following code and explain how the map works.

```
[1]: x_square = map(lambda i: i ** 2, range(1,10000000000))
```

```
[2]: for n in x_square:
         if n > 6**2:
             break
         print(n)
```

```
[2]: 1
     4
     9
     16
     25
     36
```

```
[3]: for n in x_square:
         if n > 9**2:
             break
         print(n)
```

```
[3]: 64
     81
```

Chapter 5
List Comprehension and Generators

5.1 List Comprehension

List comprehension is a way of writing programs in Python which makes this programming language very attractive. With list comprehension, it is very easy to write codes, even easier to read them, and the structure rather resembles how we describe mathematical objects in mathematics. It is a wonderful way to translate problems into a computer language with ease.

Here is the first example. If we wanted to create a list of numbers from 0 to 9 we could write:

```
[1]: list(range(10))
```

```
[1]: [0, 1, 2, 3, 4, 5, 6, 7, 8, 9]
```

With list comprehension, we can write:

```
[2]: [i for i in range(10)]
```

```
[2]: [0, 1, 2, 3, 4, 5, 6, 7, 8, 9]
```

Mathematically, this is similar to how we describe the same set in mathematics

$$\{i \mid 0 \le i \le 9\}.$$

We could also generate tuples instead of lists.

```
[3]: (i for i in range(10))
```

```
[3]: (0, 1, 2, 3, 4, 5, 6, 7, 8, 9)
```

© The Author(s), under exclusive license to Springer Nature Switzerland AG 2023
R. Hazrat, *A Course in Python*, Springer Undergraduate Mathematics Series,
https://doi.org/10.1007/978-3-031-49780-3_5

As the first example, we generate the numbers $2^n - 1$, for n between 1 and 10. Mathematically, this is the set

$$\{2^n - 1 \mid 1 \leq n \leq 10\}.$$

The list comprehension follows this description.

```
[4]: [2**n - 1 for n in range(1, 11)]
```

```
[4]: [1, 3, 7, 15, 31, 63, 127, 255, 511, 1023]
```

Before we go further, let us give other alternatives for the same program via procedural programming and functional programming, respectively.

```
[5]: L = []
     for n in range(1, 11):
         L.append(2**n - 1)
     L
```

```
[5]: [1, 3, 7, 15, 31, 63, 127, 255, 511, 1023]
```

```
[6]: list(map(lambda n: 2**n - 1, range(1,11)))
```

```
[6]: [1, 3, 7, 15, 31, 63, 127, 255, 511, 1023]
```

Exercise 5.1 *Generate the list of* $(x, \cos(x))$, *for* $x = \frac{\pi}{3}i$, *where,* $\{i \mid 0 \leq i \leq 5\}$.

Solution

Mathematically, this is the set

$$\left\{\left(x, \cos\left(\frac{\pi}{3}i\right)\right) \mid 0 \leq i \leq 5\right\}.$$

Translating this directly via list comprehension (and rounding the results of cosine) we have:

```
[7]: from math import cos, pi

     [(x, round(cos(pi/3 * x), 5)) for x in range(0, 6)]
```

```
[7]: [(0, 1.0), (1, 0.5), (2, -0.5), (3, -1.0), (4, -0.5),
      (5, 0.5)]
```

5.1.1 Putting conditions on parameters

In building up codes via list comprehension, we can introduce conditions inside the list in a natural way. As a first example, we write a program to generate all numbers between 1 and 50 which are divisible by 7.

Mathematically speaking, this is the set

$$\{n \mid 0 \leq n \leq 50, \text{ and } 7 \mid n\}.$$

In Python, using list comprehension, we can create this list, in a similar manner

```
[8]:  [n for n in range(0, 51) if n % 7 == 0]
```

```
[8]:  [0, 7, 14, 21, 28, 35, 42, 49]
```

Building on the previous code, here is a list of numbers between 1 and 199 which are divisible by 3 and by 5 but not by 7.

```
[9]:  [n for n in range(1, 200) if n % 3 == 0 and n % 5 == 0 and
       ↪not(n % 7 == 0)]
```

```
[9]:  [15, 30, 45, 60, 75, 90, 120, 135, 150, 165, 180, 195]
```

```
[10]:  [n for n in range(1, 200) if n % 15 == 0 and not(n % 7 == 0)]
```

```
[10]:  [15, 30, 45, 60, 75, 90, 120, 135, 150, 165, 180, 195]
```

Notice how naturally one can incorporate the conditions inside the list comprehension.

```
[11]:  [(n, 'even') if n % 2 == 0 else (n, 'odd') for n in
        ↪range(1,10)]
```

```
[11]:  [(1, 'odd'),
        (2, 'even'),
        (3, 'odd'),
        (4, 'even'),
        (5, 'odd'),
        (6, 'even'),
        (7, 'odd'),
        (8, 'even'),
        (9, 'odd')]
```

Exercise 5.2 *Suppose a list contains numbers and text. Write a code to extract all the text from the list.*

Solution

We create the list and then use list comprehension to collect the data.

```
[12]: x = ['Sydney', 20, 2.8, 'Brisbane', -12, [1,2,3], 'babble']
```

```
[13]: x_text = [item for item in x]; x_text
```

```
[13]: ['Sydney', 20, 2.8, 'Brisbane', -12, [1, 2, 3], 'babble']
```

```
[14]: x_text = [item for item in x if type(item) == str]
```

```
[15]: x_text
```

```
[15]: ['Sydney', 'Brisbane', 'babble']
```

```
[16]: x_rest = [item for item in x if type(item) != str]
```

```
[17]: x_rest
```

```
[17]: [20, 2.8, -12, [1, 2, 3]]
```

```
[18]: x_text + x_rest
```

```
[18]: ['Sydney', 'Brisbane', 'babble', 20, 2.8, -12, [1, 2, 3]]
```

Recall the notion of Pythagorean pairs (m, n), i.e., pairs (m, n) such that $m^2 + n^2$ is a square number. We use list comprehension to find Pythagorean pairs.

```
[19]: import math

      [(m, n) for m in range(1, 10) for n in range(1, 10)
                  if math.sqrt(m**2 + n**2).is_integer()]
```

```
[19]: [(3, 4), (4, 3), (6, 8), (8, 6)]
```

As $(6, 8)$ and $(8, 6)$ are the same pair for us, in order to avoid repetition, we can start the second loop from the starting point of the first loop.

```
[20]: [(m, n) for m in range(1, 10) for n in range(m, 10)
                  if math.sqrt(m**2 + n**2).is_integer()]
```

```
[20]: [(3, 4), (6, 8)]
```

Once we have the basic code, it is easy to modify it and obtain interesting Pythagorean pairs. For example, one can change the above code so that $m = n + 10$, so we are looking for Pythagorean pairs $(m, m + 10)$.

```
[21]: [(m, n) for m in range(1, 500) for n in range(m, 500)
                   if math.sqrt(m**2 + n**2).is_integer() and n ==␣
           ↪m + 10]
```

```
[21]: [(30, 40), (200, 210)]
```

One should be aware of the order of the loops created in the list comprehension. The outer-loop is the one which should be completed first before the inner-loop runs again. The following codes will clarify the difference:

```
[22]: [(i, j) for i in range(1, 4) for j in ['a', 'b', 'c']]
```

```
[22]: [(1, 'a'),
       (1, 'b'),
       (1, 'c'),
       (2, 'a'),
       (2, 'b'),
       (2, 'c'),
       (3, 'a'),
       (3, 'b'),
       (3, 'c')]
```

The above code should be compared with the following.

```
[23]: L = []
      for i in range(1, 4):
          for j in ['a', 'b', 'c']:
              L.append((i, j))
      L
```

```
[23]: [(1, 'a'),
       (1, 'b'),
       (1, 'c'),
       (2, 'a'),
       (2, 'b'),
       (2, 'c'),
       (3, 'a'),
       (3, 'b'),
       (3, 'c')]
```

Exercise 5.3 *Show that the only n less than* 1000 *such that*

$$3^n + 4^n + \cdots + (n+2)^n = (n+3)^n$$

are the numbers 2 and 3.

Solution

Recall the function `sum` gives the sum of the entries of a list.

```
[24]: for n in range(1, 1001):
          x = [i**n for i in range(3, n+3)]
          if sum(x) == (n + 3)**n:
              print(n)
```

[24]: 2
      3

We can incorporate the above code into one long (nested) list comprehension. We leave it to the reader to judge which code is easier to read and understand.

```
[25]: [n for n in range(1, 1001) if
                           sum([i**n for i in range(3, n+3)])↵
       ↵== (n + 3)**n]
```

[25]: [2, 3]

Exercise 5.4 *Recall the Collatz function*

$$f(x) = \begin{cases} x/2 & \text{if } x \text{ is even} \\ 3x + 1 & \text{if } x \text{ is odd.} \end{cases}$$

and produce $f(n), 1 \leq n < 20$, *using list comprehension.*

Solution

Here is the code:

```
[26]: [n//2 if n % 2 == 0 else 3 * n + 1 for n in range(1, 20)]
```

[26]: [4, 1, 10, 2, 16, 3, 22, 4, 28, 5, 34, 6, 40, 7, 46, 8, 52,↵
       ↵9, 58]

Exercise 5.5 *Find the first 5 positive integers n such that* $n^6 + 1091$ *is prime. Show that all these n are between* 3500 *and* 8500.

Solution

We will be using the `isprime` function available in `sympy` to test the numbers.

```
[27]: from sympy import isprime
```

Since the question gives a clue how far one needs to go, we create a list `range(1,10000)` and check among n in this list when $n^6 + 1091$ is prime.

```
[28]: [n for n in range(1, 10000) if isprime(n**6 + 1091)]
```

```
[28]: [3906, 4620, 5166, 5376, 5460, 8190]
```

If one didn't know how far one needs to go to get all five positive n such that $n^6 + 1091$ is prime, one option would have been to create a `while` loop.

```
[29]: count = 0
      n = 0
      while count < 5:
          n += 1
          if isprime(n**6 + 1091):
              print(n)
              count += 1
```

```
[29]: 3906
      4620
      5166
      5376
      5460
```

Exercise 5.6 *Determine all the positive integers n between 3 and 50 for which* 2^{2008} *is divisible by*

$$1 + \binom{n}{1} + \binom{n}{2} + \binom{n}{3}.$$

Solution

Recall that the binomial coefficient (the number of ways of choosing m-items among n-items) is

$$\binom{n}{m} := \frac{n!}{m!(n-m)!}.$$

Both the factorial and the binomial coefficient `comb` are available in the `math` library.

```
[30]: from math import comb, factorial
```

We first check the method `comb` against the definition!

```
[31]: comb(5,3) == factorial(5) // (factorial(3) * factorial(5 -⌴
      ↪3))
```

```
[31]:  True
```

```
[32]:  [n for n in range(3, 51) if 2**2008 % (1 + comb(n,1) +↵
         ↪comb(n,2) + comb(n,3)) == 0]
```

```
[32]:  [3, 7, 23]
```

We now write Exercise 1.6 with list comprehension.

Exercise 5.7 *Let m be a natural number and*

$$A = \frac{(m+3)^3 + 1}{3m}.$$

Find all the integers m less than 500 such that A is an integer. Show that A is always odd.

Solution

We define the function $A(m)$ and then compose a list comprehension to find m's such that $A(m)$ is an integer

```
[33]:  def A(m):
           return ((m + 3)**3 +1)/(3*m)


       [m for m in range(1,500) if A(m).is_integer()]
```

```
[33]:  [2, 14]
```

Here is yet another way to write the code using functional programming.

```
[34]:  def A(m):
           A = ((m + 3)**3 +1)/(3*m)
           return A.is_integer()


       list(filter(A, range(1, 500)))
```

```
[34]:  [2, 14]
```

Exercise 5.8 *Explain what the following code does:*

```
[35]: import math

      isprime = lambda x: all(x % k for k in range(2, 1 + math.
        isqrt(x)))
      [n for n in range(350, 400) if isprime(n**2 + n + 1)]
```

```
[35]: [351, 357, 369, 378, 381, 383, 392, 395, 398]
```

Solution

The `all()` function returns True if all items in an iterable are true, otherwise it returns False. The tuple comprehension `(x % k for k in range(2, 1 + math.isqrt(x))` checks if the number x is divisible by numbers up to $\sqrt{x}$. If there is an instance when x is indeed divisible, then `x % k` returns 0. Within Boolean statements, all positive numbers represent True and zero represent False. Thus if there is one number dividing x, using `all` the combinations of Boolean outputs inside the tuple comprehension is False, otherwise the result is True meaning no number divides x, i.e., x is a prime number. The next list comprehension in the code uses this prime test to find prime numbers of the form of n^2+n+1 for $350 \leq n \leq 400$.

5.2 Sets and Dictionaries

5.2.1 Sets

To handle a collection of data, we have worked with `list` and `tuple`. With `list` we can collect data, access the elements, and modify the list by adding or deleting items, i.e. they are mutable objects. `tuples` are immutable but we can still access their elements.

Next we introduce two more objects for handling data, `sets` and `dictionaries`. The concept of `sets` in Python is the closest to the notion of sets in mathematics: a collection of data where order and duplications of members does not change the set.

```
[36]: X = {1, 2, 3, 2, 1}
```

```
[37]: X
```

```
[37]: {1, 2, 3}
```

```
[38]: {1, 2} == {2, 1}
```

[38]: True

[39]: `{1, 1, 2} == {1, 2}`

[39]: True

Compare the above with the examples of lists in Chapter 2, where we get False in all cases.

Mathematical operations on sets such as union, intersection, and complements are available within set objects. We are going to demonstrate these methods with two sets; the set of prime numbers smaller than 20 and the set of odd numbers up to 20.

[40]: `X = {2, 3, 5, 7, 11, 13, 17, 19}`

[41]: `Y = {3, 5, 7, 9, 11, 13, 15, 17, 19}`

Recall the operations of the union and intersection of two sets:

$$X \cup Y = \{x \mid x \in A \text{ or } x \in B\}$$

$$X \cap Y = \{x \mid x \in A \text{ and } x \in B\}.$$

[42]: `X.union(Y)`

[42]: `{2, 3, 5, 7, 9, 11, 13, 15, 17, 19}`

[43]: `X.union(['a','b'])`

[43]: `{11, 13, 17, 19, 2, 3, 5, 7, 'a', 'b'}`

Note that union is a method belonging to the set object. However this method does not change the object.

[44]: `X`

[44]: `{2, 3, 5, 7, 11, 13, 17, 19}`

[45]: `X | Y == X.union(Y)`

[45]: True

[46]: `X & Y == X.intersection(Y)`

[46]: True

The operation

$$X \setminus Y = \{x \in X | x \notin Y\}$$

can be obtained in Python by using '.difference'

```
[47]: X - Y
```

[47]: {2}

```
[48]: X.difference(Y)
```

[48]: {2}

```
[49]: X
```

[49]: {2, 3, 5, 7, 11, 13, 17, 19}

The operation of symmetric difference, namely

$$X \setminus Y \bigcup Y \setminus X,$$

sometimes denoted by $X \Delta Y$ in mathematics books, can be achieved by symmetric_difference.

```
[50]: X.symmetric_difference(Y)
```

[50]: {2, 9, 15}

```
[51]: X^Y
```

[51]: {2, 9, 15}

```
[52]: Y - X
```

[52]: {9, 15}

```
[53]: X - Y
```

[53]: {2}

```
[54]: (Y-X).union(X-Y) == X^Y
```

[54]: True

For the empty set, use set() as { } is reserved for the empty dictionary, which we will study later.

[55]: `X = set(); X`

[55]: `set()`

[56]: `X.union({2})`

[56]: `{2}`

Note that one cannot access items in a set by referring to an index (as we do for lists or tuples), but we can loop through the set items using a for loop, or ask if a specified value is present in a set, by using the in keyword.

Exercise 5.9 *Find the set of remainders of $n^2 + n + 41$ upon dividing by 7 for $0 \le n < 100$.*

Solution

We will use a set to collect the remainders as we are only interested in the actual remainders and not how many of them we obtain when dividing $n^2 + 41$ by 7.

[57]:
```
A = set()
for n in range(100):
    A = A | {(n**2 + n + 41) % 7}
print(A)
```

[57]: `{1, 4, 5, 6}`

Recall the shorthand a #= b, which is equivalent to a = a # b.

[58]:
```
A = set()
for n in range(100):
    A |= {(n**2 + n + 41) % 7}
print(A)
```

[58]: `{1, 4, 5, 6}`

[59]:
```
A = set()
for n in range(100):
    A = A.union({(n**2 + n + 41) % 7})
print(A)
```

[59]: `{1, 4, 5, 6}`

Recall the list comprehension approach to programming. We can use "set comprehension" to do the job.

```
[60]: A=[(n**2 + n + 41) % 7 for n in range(101)]
```

```
[61]: print(A)
```

```
[61]: [6, 1, 5, 4, 5, 1, 6, 6, 1, 5, 4, 5, 1, 6, 6, 1, 5, 4, 5, 1,
       6, 6, 1, 5, 4, 5, 1, 6, 6, 1, 5, 4, 5, 1, 6, 6, 1, 5, 4, 5,
       1, 6, 6, 1, 5, 4, 5, 1, 6, 6, 1, 5, 4, 5, 1, 6, 6, 1, 5, 4,
       5, 1, 6, 6, 1, 5, 4, 5, 1, 6, 6, 1, 5, 4, 5, 1, 6, 6, 1, 5,
       4, 5, 1, 6, 6, 1, 5, 4, 5, 1, 6, 6, 1, 5, 4, 5, 1, 6, 6, 1,
       5]
```

```
[62]: B={(n**2 + n + 41) % 7 for n in range(101)}
```

```
[63]: print(B)
```

```
[63]: {1, 4, 5, 6}
```

Exercise 5.10 *Consider the following sets U, A and B and find the shaded areas in the pictures.*

```
U = {-34, 23, 50, 'cat', 'dog', 'mouse', 'food',
'one','sydney'}
A = {'cat', 'dog', 50, 'sydney', 'food'}
B = {'dog', 23, -34, 50, 'food'}
```

```
[64]: from PIL import Image
      Image.open("setpic.png")
```

[64]:

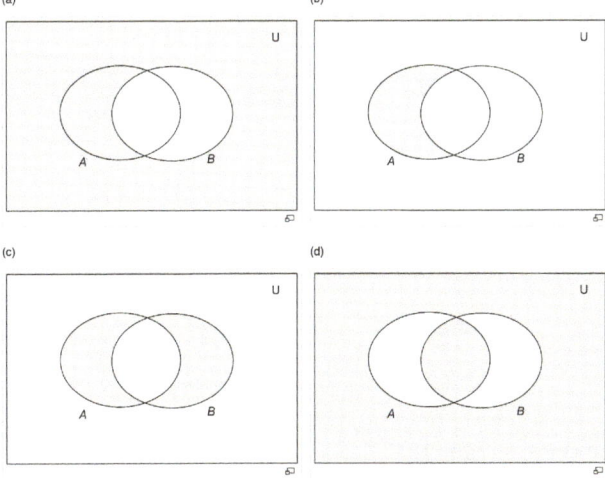

Solution

We first define these sets.

```
[65]: U = {-34, 23, 50, 'cat', 'dog', 'mouse', 'food',↵
       ↪'one','sydney'}
```

```
[66]: A = {'cat', 'dog', 50, 'sydney', 'food'}
```

```
[67]: B = {'dog', 23, -34, 50, 'food'}
```

To obtain (a):

```
[68]: U - B
```

```
[68]: {'cat', 'mouse', 'one', 'sydney'}
```

To obtain (b):

```
[69]: A - B
```

```
[69]: {'cat', 'sydney'}
```

To obtain (c), we can do one of the following:

```
[70]: (A - B) | (B - A)
```

```
[70]: {-34, 23, 'cat', 'sydney'}
```

```
[71]: A.symmetric_difference(B)
```

```
[71]: {-34, 23, 'cat', 'sydney'}
```

```
[72]: B.symmetric_difference(A)
```

```
[72]: {-34, 23, 'cat', 'sydney'}
```

To obtain (d):

```
[73]: A^B
```

```
[73]: {-34, 23, 'cat', 'sydney'}
```

```
[74]: B^A
```

```
[74]: {-34, 23, 'cat', 'sydney'}
```

[75]: U - (A^B)

[75]: {50, 'dog', 'food', 'mouse', 'one'}

[76]: U - A.symmetric_difference(B)

[76]: {50, 'dog', 'food', 'mouse', 'one'}

Exercise 5.11 *Let U be a set and A a subset of U. Then the* complement *of A is defined as U\A and denoted by A^c. One can prove that for subsets $A, B \subseteq U$*

$$(A \cup B)^c = A^c \cap B^c$$

Demonstrate this identity with the example above.

Solution

We only need to translate this to Python:

[77]: U - (A | B) == (U - A) & (U - B)

[77]: True

Exercise 5.12 *Demonstrate the following identities with the example above*

$$A \cap (A \cup B) = A$$

$$(A \cap B) \cup (A \cup B^c)^c = B$$

Solution

Again, we need to translate these identities into Python:

[78]: A & (A | B) == A

[78]: True

[79]: (A & B) | (U - (A | (U - B))) == B

[79]: True

Exercise 5.13 *How many different words are used in the The Book of Genesis! Find the longest words in the book.*

Solution

Recall the library `nltk`. We upload the library and get the entire text of Genesis. Then using `set` we get rid of the repeated words.

```
[80]: import nltk
      from nltk.book import text3
```

```
[80]: *** Introductory Examples for the NLTK Book ***
      Loading text1, ..., text9 and sent1, ..., sent9
      Type the name of the text or sentence to view it.
      Type: 'texts()' or 'sents()' to list the materials.
      text1: Moby Dick by Herman Melville 1851
      text2: Sense and Sensibility by Jane Austen 1811
      text3: The Book of Genesis
      text4: Inaugural Address Corpus
      text5: Chat Corpus
      text6: Monty Python and the Holy Grail
      text7: Wall Street Journal
      text8: Personals Corpus
      text9: The Man Who Was Thursday by G . K . Chesterton 1908
```

```
[81]: text3
```

```
[81]: <Text: The Book of Genesis>
```

```
[82]: len(text3)
```

```
[82]: 44764
```

There are about 44 thousand words. But as this is a list, many of the words have been repeated. Using `set` we can drop the repeated words.

```
[83]: g_words= set(text3)
```

```
[84]: len(g_words)
```

```
[84]: 2789
```

```
[85]: {len(word) for word in g_words}
```

```
[85]: {1, 2, 3, 4, 5, 6, 7, 8, 9, 10, 11, 12, 13, 14, 15}
```

```
[86]:  {word for word in g_words if len(word)==15}
```

```
[86]:  {'Zaphnathpaaneah', 'interpretations'}
```

5.2.2 Dictionaries

Dictionaries are another way to collect and organise data. They come in pairs, a key and values associated to the key. Both keys and values are objects. However keys are unmutable objects whereas values can be mutable. We start by defining a dictionary which consists of four keys, A, B, C, D and the values associated to them. The example shows how to access the values and how to add more data to the dictionary.

```
[87]:  score = {'A': 100, 'B': 70, 'C': 50, 'F': 0 }
```

```
[88]:  score['A']
```

```
[88]:  100
```

```
[89]:  score['C']
```

```
[89]:  50
```

```
[90]:  score['G'] = 'back'
```

```
[91]:  score
```

```
[91]:  {'A': 100, 'B': 70, 'C': 50, 'F': 0, 'G': 'back'}
```

```
[92]:  score['m_range']=[49, 48, 47, 46]
```

```
[93]:  score
```

```
[93]:  {'A': 100, 'B': 70, 'C': 50, 'F': 0, 'G': 'back', 'm_range':␣
       ↪[49, 48, 47, 46]}
```

We can retrieve the keys and values of the dictionaries.

```
[94]:  score.keys()
```

```
[94]:  dict_keys(['A', 'B', 'C', 'F', 'G', 'm_range'])
```

```
[95]:  score.values()
```

```
[95]: dict_values([100, 70, 50, 0, 'back', [49, 48, 47, 46]])
```

```
[96]: [score[i] for i in score.keys()]
```

```
[96]: [100, 70, 50, 0, 'back', [49, 48, 47, 46]]
```

In fact, the .keys() is not needed above, since iterating over a mapping is the same as iterating over its keys.

```
[97]: [score[i] for i in score]
```

```
[97]: [100, 70, 50, 0, 'back', [49, 48, 47, 46]]
```

Similar to lists and tuples, we can check if a key belongs to a dictionary.

```
[98]: 'A' in score
```

```
[98]: True
```

```
[99]: 70 in score
```

```
[99]: False
```

```
[100]: del score['F']
```

```
[101]: score
```

```
[101]: {'A': 100, 'B': 70, 'C': 50, 'G': 'back', 'm_range': [49, 48,↵
        ↪47, 46]}
```

```
[102]: score.pop('m_range')
```

```
[102]: [49, 48, 47, 46]
```

```
[103]: score
```

```
[103]: {'A': 100, 'B': 70, 'C': 50, 'G': 'back'}
```

```
[104]: score.update({'check':34, 'uncheck':-30})
```

```
[105]: score
```

```
[105]: {'A': 100, 'B': 70, 'C': 50, 'G': 'back', 'check': 34,↵
        ↪'uncheck': -30}
```

One way to build a dictionary from two lists is to use the method `zip`. If $x = (x_1, x_2, \ldots, x_n)$ and $y = (y_1, y_2, \ldots, y_n)$, then `zip(x,y)` will pair elements of x and y.

```
[106]: x = ('a', 'b', 'c')
       y = (1, 2, 3)
```

```
[107]: list(zip(x, y))
```

```
[107]: [('a', 1), ('b', 2), ('c', 3)]
```

Now we can build a dictionary. Here are three different ways to do this:

```
[108]: L={}
       for key, value in zip(x,y):
           L[key] = value

       L
```

```
[108]: {'a': 1, 'b': 2, 'c': 3}
```

```
[109]: {key: value for key, value in zip(x,y)}
```

```
[109]: {'a': 1, 'b': 2, 'c': 3}
```

The easiest way is to use `dict`

```
[110]: L = dict(zip(x, y))
       L
```

```
[110]: {'a': 1, 'b': 2, 'c': 3}
```

Oftentimes we need to return a certain value if the key we are looking for is not in the list. In this case we can use `get` with a default value.

```
[111]: L
```

```
[111]: {'a': 1, 'b': 2, 'c': 3}
```

```
[112]: L['b']
```

```
[112]: 2
```

The input `L['z']` would generate an error.

```
[113]: L.get('b')
```

[113]: 2

[114]: L.get('z','does not exist')

[114]: 'does not exist'

[115]: L

[115]: {'a': 1, 'b': 2, 'c': 3}

Finally, once we get access to a value in the dictionary, we can use all the methods available to that particular object.

[116]: L['a'] = ['Momento']

[117]: L

[117]: {'a': ['Momento'], 'b': 2, 'c': 3}

[118]: L['a'].append('Mori')

[119]: L

[119]: {'a': ['Momento', 'Mori'], 'b': 2, 'c': 3}

Exercise 5.14 *Write a program to create a dictionary with two keys* even *and* odd *and collect even and odd numbers from 1 to 20 into the corresponding keys.*

Solution

First, we define a dictionary, consisting of two keys, even and odd, each having the value of an empty list. We then go through the numbers adding even numbers to the even key and odd numbers to the odd key.

```python
[120]:  L = {'even':[], 'odd':[]}
        for i in range(21):
            if i%2:
                L['odd'].append(i)
            else:
                L['even'].append(i)
        print(L)
```

[120]: {'even': [0, 2, 4, 6, 8, 10, 12, 14, 16, 18, 20],
         'odd': [1, 3, 5, 7, 9, 11, 13, 15, 17, 19]}

Here is another clever way to write the same code.

```
[121]:  L = {'even':[], 'odd':[]}
        for i in range(21):
            L[('even', 'odd')[i%2]].append(i)
        print(L)
```

```
[121]:  {'even': [0, 2, 4, 6, 8, 10, 12, 14, 16, 18, 20],
         'odd': [1, 3, 5, 7, 9, 11, 13, 15, 17, 19]}
```

Godfrey Hardy, a very prominent English mathematician who was also a bit eccentric, invited the Indian mathematician Srinivasa Ramanujan to go to Cambridge. Ramanujan became an exceptional collaborator of Hardy. Hardy recalled:

I remember once going to see him when he was ill at Putney. I had ridden in taxi cab number 1729 and remarked that the number seemed to me rather a dull one, and that I hoped it was not an unfavourable omen. "No," Ramanujan replied, "it is a very interesting number; it is the smallest number expressible as the sum of two cubes in two different ways.

Exercise 5.15 *Write a code to find the smallest number expressible as the sum of two cubes in two different ways.*

Solution

We will be using `dictionary` to find the smallest number. The code consists of two `dictionaries`: `cube` and `rama`. In cube the keys are positive integers and the values are their cubes. We add to the cube more keys and their values as long as needed, until we find the desired number. For `rama`, the keys are the sums of cubes $b^3 + a^3$ and the associated value is (b, a). We generate the dictionary `rama` as follows:

$$\{0^3 + 0^3 : (0,0),$$
$$0^3 + 1^3 : (0,1),$$
$$0^3 + 2^3 : (0,2), 1^3 + 2^3 : (1,2),$$
$$0^3 + 3^3 : (0,3), 1^3 + 3^3 : (1,3), 2^3 + 3^3 : (2,3), ...\}$$

Now each time we create a new sum of cubes $y^3 + x^3$, we first check if this has already appeared in `rama` by asking `cube(y)+cube(x) in rama`. If this is not the case then we add it to `rama`, namely `rama[cube(y)+cube(x)]=(y,x)`. However, if it is the case that `cube(y)+cube(x)` is already in `rama`, then there is a key `cube(b)+cube(a)` which matches this. Thus $y^3 + x^3 = b^3 + a^3$. In this case we can print both the key and also the pairs (y,x) and (b,a) and we have completed the code.

Here is the actual code:

```
[122]: rama, cube, a = {}, {}, 0
       cond = True
       while cond:
           cube[a] = a * a * a
           for b in range(a):
               s = cube[b] + cube[a]
               if s in rama:
                   print((s, (b, a), rama[s]))
                   cond = False
               rama[s] = (b ,a)
           a += 1
```

```
[122]: (1729, (1, 12), (9, 10))
```

As an further exercise, modify the above code to find more numbers that can be expressible as the sum of two cubes in two different ways.

5.3 Generators

A generator is one of those tools in Python that allows us to write codes that can handle a very large number of items, even when we don't know a priori how many items we are dealing with. In fact we can handle infinity itself!

When we define a list, the sequence of items in the list are all already cooked and ready and can be used right away, whereas with generators, the items in the sequence will be, well, generated upon request. The following examples make this clear:

```
[123]: l = [i**5 for i in range(10)]
```

```
[124]: l
```

```
[124]: [0, 1, 32, 243, 1024, 3125, 7776, 16807, 32768, 59049]
```

All the elements of the list l have now been generated. l contains 10 elements. l can be used as many times as one would like. Now as for a generator:

```
[125]: g = (i**5 for i in range(10000000))
```

Notice the difference between the definition of l and g. In g, where the range goes all the way up to 10 million, Python does not create the list of 10 million numbers! (it might not even make sense to do so!) The definition of g is there and when it is needed we can prompt Python to generate as many elements of g as needed.

```
[126]: [next(g) for _ in range(10)]
```

```
[126]:  [0, 1, 32, 243, 1024, 3125, 7776, 16807, 32768, 59049]
```

The function `next` prompts Python to generate the next item of the generator. It also keeps track of where the item in the sequence is located. Therefore if we call `next(g)` ten times, it will create the first 10 numbers in the sequence and stop there. The beauty of this is, Python still knows where the position of the *iterator* is, so next time when we prompt g, it starts from there.

```
[127]:  [next(g) for _ in range(5)]
```

```
[127]:  [100000, 161051, 248832, 371293, 537824]
```

```
[128]:  [next(g) for _ in range(5)]
```

```
[128]:  [759375, 1048576, 1419857, 1889568, 2476099]
```

Exercise 5.16 *Find the first 5 positive integers n such that $n^6 + 1091$ is prime.*

Solution

We have seen this exercise before. The exercise previously also told us that the first five n's such that $n^6 + 1091$ is prime are between 3500 and 8500. Therefore we knew from the outset how far we needed to increase n. If we didn't know the bound for n, then the notion of generators allows us to write a one-line code for this exercise. Since the generators run the code upon request, we can define a tuple comprehension, allowing for n to go all the way to a very large number, say 1 million, checking if $n^6 + 1091$ is prime. Once this generator is defined, we prompt it five times, so it will give us the first five n, as wanted.

```
[129]:  from sympy import isprime

        x = (n for n in range(1,100000) if isprime(n**6 + 1091))
```

Note that at the moment `x` is a generator and we need to prompt it so that we get the first n such that $n^6 + 1091$ is prime.

```
[130]:  [next(x) for i in range(5)]
```

```
[130]:  [3906, 4620, 5166, 5376, 5460]
```

In order to obtain the first five items in one go, we could use a method from the `itertools` library.

```
[131]:  import itertools

        list(itertools.repeat(next(x), 5))
```

[131]: [8190, 8190, 8190, 8190, 8190]

Or we could use another method, also available in the wonderful itertools library.

[132]:
```python
import itertools

list(itertools.islice(x, 5))
```

[132]: [13020, 13986, 14490, 17934, 19740]

Note that since the pointer was at the fifth prime, when we prompt the generator x again, it would give us the next 5 primes generated by the formula $n^6 + 1091$.

Exercise 5.17 *Find the smallest multiple of 99999 that contains no 9's amongst its digits.*

Solution

We have seen this exercise before, where we used a while-loop to get the answer. Here we use the concept of generators to approach the problem.

[133]:
```python
import itertools

next(x for x in (99999 * k for k in itertools.count(1)) if
    ↵'9' not in str(x))
```

[133]: 1111188888

Here the entire line

```python
(x for x in (99999 * k for k in itertools.count(1)) if '9' not
in str(x))
```

is a generator which generates a multiple of 99999 that does not contain the digit 9. Each time we prompt it, it gives us the first instance it arrives at.

[134]:
```python
t = (x for x in (99999 * k for k in itertools.count(1)) if
    ↵'9' not in str(x))

[next(t) for i in range(3)]
```

[134]: [1111188888, 1111288887, 1111388886]

The library itertools provides very interesting methods to compose codes with. We will look at one of them, takewhile, and encourage the reader to explore the others.

```
[135]: import nltk
       from nltk.book import text3
```

```
[136]: words = list(itertools.takewhile(lambda x: x != 'morning',␣
       ↪text3))
```

```
[137]: " ".join(words)
```

```
[137]: 'In the beginning God created the heaven and the earth . And
       the earth was without form , and void ; and darkness was upon
       the face of the deep . And the Spirit of God moved upon the
       face of the waters . And God said , Let there be light : and
       there was light . And God saw the light , that it was good :
       and God divided the light from the darkness . And God called
       the light Day , and the darkness he called Night . And the
       evening and the'
```

The method `takewhile(cond, iter)` goes through the `iter` list until the condition cond fails.

5.3.1 Generator functions

We can define a generator via functions, the so-called *generator functions*. We describe the idea with an example. Recall the function which returns the *n*-th Fibonacci number.

```
[138]: def fibs(n):
           x, y = 0, 1
           for _ in range(n):
               x, y = y, x+y
           return x
```

```
[139]: fibs(10)
```

```
[139]: 55
```

We modify it as a generator as follows:

```
[140]: def gfibs(n):
           x, y = 0, 1
           for _ in range(n):
               x, y = y, x+y
               yield x
```

Notice that in the generator function, we have replaced `return` by `yield` and the `yield` is located within the loop. There is a substantial difference between `fibs` and `gfibs`. When we run `fibs(1000)` Python calculates the 1000-th Fibonacci number and returns it, whereas `gfibs(1000)` just defines such a generator. Each time we call this function, the variable _ runs one time inside the loop, the next Fibonacci number in the sequence is generated, and `yield` keeps this value and the generator stops there. If we call the function again, we will get the next number in the list. So the code `fibs(1000000)` might mess up the computer memory whereas `gfibs(1000000)` is in standby and we can access the Fibonacci numbers up to 1000000 when needed.

```
[141]:  d = gfibs(10000000)
```

```
[142]:  [next(d) for _ in range(10)]
```

```
[142]:  [1, 1, 2, 3, 5, 8, 13, 21, 34, 55]
```

```
[143]:  [next(d) for _ in range(10)]
```

```
[143]:  [89, 144, 233, 377, 610, 987, 1597, 2584, 4181, 6765]
```

Here is an amusing example showing how the generator functions keeps track of each call.

```
[144]:  def gene(n):
            count = 0
            while count < n:
                count += 1
                yield count, 'loop' * count

        for i in gene(5):
            print(i)
```

```
[144]:  (1, 'loop')
        (2, 'looploop')
        (3, 'looploopfloop')
        (4, 'looploopflooploop')
        (5, 'loopfloopflooplooploop')
```

We revisit Exercise 4.8, which we can now write using generators.

Exercise 5.18 *Define a sequence of numbers by $a_0 = 1$, $a_1 = 1$ and, for $n \geq 2$, $a_n = 3a_{n-1} - a_{n-2}$. Write a function to accept n and calculate a_n, for $0 \leq n \leq 10$. function to accept n and calculate a_n, for $0 \leq n \leq 10$.*

Solution

Here is the code, defining a function with 'yield', which gives a generator.

```
[145]: import itertools

       def aseq(a, b):
           while True:
               s = 3 * a - b
               a, b = s, a
               yield b

       list(itertools.islice(aseq(1, 1), 10))
       [1, 1, 2, 5, 13, 34, 89, 233, 610, 1597]
```

```
[145]: [1, 1, 2, 5, 13, 34, 89, 233, 610, 1597, 4181]
```

Exercise 5.19 *About 110 years ago, in 1914, Ramanujan came up with the following formula for $1/\pi$. He always said the ideas for these types of formulas were divinely inspired, and a glance at the formula makes one believe it!*

$$\frac{1}{\pi} = \frac{\sqrt{8}}{99^2} \sum_{n=0}^{\infty} \frac{(4n)!}{(4^n n!)^4} \frac{1103 + 26390n}{99^{4n}}$$

Write a program to see how many terms of the sum in the right-hand side are needed to approximate $1/\pi$ up to 17-significant digits.

Solution

Notice that Ramanujan's formula consists of an infinite sum. Of course we can't let a program loop infinitely many times, but what we can do is go as far as needed to get the desired approximation. The generator count does just that.

```
146]: from itertools import count
```

```
147]: from math import factorial, sqrt, pi

      def Ram1914(n):
          a = (1103 + 26390 * n) / (99**(4 * n))
          b = factorial(4 * n)/(4**n * factorial(n))**4
```

```
    return a * b

s = 0
c = sqrt(8) / 99**2

for i in count():
    s += Ram1914(i)
    if abs(c * s  - 1 / pi) < 0.00000000000000001:
        print(f'Number of iterations: {i+1}, Ramanujan↵
↳formula: {c * s} and 1/pi: {1/pi}')
        break
```

[147]: Number of iterations: 3, Ramanujan formula: 0.3183098861837907
        and 1/pi: 0.3183098861837907

It is remarkable that with just three iterations we can get so close to $1/\pi$.

Problems

1. Investigate whether the following identity holds:

$$(1 + 2 + 3 + \cdots + n)^2 = (1^3 + 2^3 + 3^3 + \cdots + n^3).$$

Note: Since we haven't yet worked with symbolic computations in Python, one approach is to check that the above identity holds for various values for n.

2. Consider the following series:

$$\frac{1 \times 3}{2 \times 2} \frac{3 \times 5}{4 \times 4} \frac{5 \times 7}{6 \times 6} \cdots .$$

Investigate if this series tends to $2/\pi$.

Note: The general term of this series is

$$\frac{(2n - 1)(2n + 1)}{2n \times 2n}.$$

3. Given non-negative integers $c_1, c_2, \ldots, c_m$, with $c_m \neq 0$, define the function

$$f(c_1, c_2, \ldots, c_m) = c_1 - \cfrac{1}{c_2 - \cfrac{2}{\cdots - \cfrac{m-1}{c_m}}}.$$

Note that if a denominator happens to become zero, the program should give a warning and stop the execution.

4. Investigate if

$$\frac{2}{\pi} = \frac{\sqrt{2}}{2} \frac{\sqrt{2 + \sqrt{2}}}{2} \frac{\sqrt{2 + \sqrt{2 + \sqrt{2}}}}{2} \cdots$$

5. An integer $d_n d_{n-1} d_{n-2} \ldots d_1$ is *palindromic* if

$$d_n d_{n-1} d_{n-2} \ldots d_1 = d_1 d_2 \ldots d_{n-1} d_n,$$

(for example 15651). Write a code to ask for a number $d_n d_{n-1} d_{n-2} \ldots d_1$ and find out if it is palindromic. Enhance the code further so that if the number is not palindromic then the code tests whether $d_n d_{n-1} d_{n-2} \ldots d_1 + d_1 d_2 \ldots d_{n-1} d_n$ is (for example, 108+801=909). Furthermore, write a code to give the number of times needed to repeat this procedure until one gets a palindromic number, starting with $d_n d_{n-1} d_{n-2} \ldots d_1$ (if it takes more than 150 times, let the function return infinity).

6. Write a code to check for every integer n that the following inequality holds:

$$\sqrt{2} \sqrt[4]{4} \sqrt[8]{8} \cdots \sqrt[2n]{2n} \le n + 1.$$

7. A *happy* number is a number such that if one squares its digits and adds them together, and then takes the result and squares its digits and adds them together again, and so on, repeating this process, then one eventually reaches the number 1. Find all the happy ages, i.e., happy numbers up to 100.

8. Consider a positive integer. Sort the decimal digits of this number in ascending and descending order. Calculate the difference of these two numbers (for example, starting from 5742, we get 7542 − 2457 = 5085). This is called the Kaprekar routine. First check that starting with any 4-digit number and repeating the Kaprekar routine, you always reach either 0 or 6174. Then find out, among all the 4-digit numbers, what is the maximum number of iterations needed in order to get to 6174.

9. Explain what the following code does:

```
[1]: [n//2 if n % 2 == 0 else 3*n + 1 for n in range(1,30) if
     ↪n % 3 == 1]
```

```
[1]: [4, 2, 22, 5, 40, 8, 58, 11, 76, 14]
```

10. Explain what the following code does:

```
[1]:  m = [('a', 'b', 'c'), ('d', 'e', 'f'), ('g', 'h', 'k')]
```

```
[2]:   [list(i) for i in zip(*m)]
```

11. We asked ChatGPT to write a code to find Ramanujan's cab number, namely
    the smallest positive integer that can be written as a sum of two cubes in two
    different way. The code given is as follows. It does not generate the result, i.e.,
    1729. Where does the code go wrong?

```
[1]:   def find_smallest_sum_of_two_cubed():
           # Initialize the smallest sum to a large value
           smallest_sum = float('inf')

           # Define the upper limit for the search (You can
       ↪adjust this if needed)
           upper_limit = 1000

           # Dictionary to store the sums of cubes and their
       ↪corresponding numbers
           sums_of_cubes = {}

           for num in range(1, upper_limit):
               cube = num**3

               for key in sums_of_cubes:
                   current_sum = key + cube

                   if current_sum in sums_of_cubes:
                       # Check if the numbers are different to
       ↪find two different ways
                       if sums_of_cubes[current_sum] != num:
                           # Update the smallest sum if a
       ↪smaller one is found
                           if current_sum < smallest_sum:
                               smallest_sum = current_sum
                   else:
                       sums_of_cubes[current_sum] = num

           return smallest_sum

       result = find_smallest_sum_of_two_cubed()
       print(f"The smallest number that can be written as a sum
       ↪of two cubes in two different ways is: {result}")
```

```
[1]:   The smallest number that can be written as a sum of two
       cubes in two different ways is: inf
```

Chapter 6
The sympy Library

6.1 sympy, Symbolic Python

One of the abilities of Python, via its library sympy, is to handle symbolic computations, i.e., Python can comfortably work with symbols (we have seen some examples of this already in Chapter 1). Working symbolically allows one to treat the situation "abstractly" without assigning any value to the parameters. When it is needed, one can then assign specific objects to the parameters. The sympy library also allows us to do calculus symbolically, solve equations and calculate derivatives and integrals. In this chapter we will look at some of these features.

As a first example, consider the expression $(x + y)^2$. One can use Python to expand this expression symbolically. All we need to do is to import the library sympy, introduce the symbols x and y to the program, and then use the method expand.

```
[1]: from sympy import symbols, factor, expand
```

```
[2]: x = symbols('x')
     y = symbols('y')
```

```
[3]: x + y
```

[3]: $x + y$

One can also introduce a sequence of symbols

```
[4]: x, y = symbols('x y')
```

```
[5]: x**2 + y**2
```

[5]: $x^2 + y^2$

As one can see, Python can now handle the symbols x and y.

© The Author(s), under exclusive license to Springer Nature Switzerland AG 2023
R. Hazrat, *A Course in Python*, Springer Undergraduate Mathematics Series,
https://doi.org/10.1007/978-3-031-49780-3_6

[6]: `(x + y)**2`

[6]: $(x + y)^2$

[7]: `expand((x + y)**2)`

[7]: $x^2 + 2xy + y^2$

[8]: `expand((x - y)**5)`

[8]: $x^5 - 5x^4y + 10x^3y^2 - 10x^2y^3 + 5xy^4 - y^5$

As one might expect, Python can handle more involved polynomial arithmetic. We are going to expand

$$\frac{(x + y)(3x - 6y)^4}{(2x - y)^2}.$$

[9]: `expand((x + y) * (3*x - 6*y)**4 / (2*x - y)**2)`

[9]: $\dfrac{81x^5}{4x^2 - 4xy + y^2} - \dfrac{567x^4y}{4x^2 - 4xy + y^2} + \dfrac{1296x^3y^2}{4x^2 - 4xy + y^2} - \dfrac{648x^2y^3}{4x^2 - 4xy + y^2} - \dfrac{1296xy^4}{4x^2 - 4xy + y^2} + \dfrac{1296y^5}{4x^2 - 4xy + y^2}$

The method `factor` can do the inverse of `expand`, namely factorise an expression.

[10]: `factor(x**3 - y**3)`

[10]: $(x - y)\left(x^2 + xy + y^2\right)$

While expansion of an algebraic expression is a simple and routine procedure, the factorisation of algebraic expressions is often quite challenging. My favourite example is this one: Try to factorise the expression $x^{10} + x^5 + 1$. Here is one way to do it:

$x^{10} + x^5 + 1$ (adding $x^i - x^i$, $1 \leq i \leq 9$, to the expression we have)

$= x^{10} + \underbrace{x^9 - x^9} + \underbrace{x^8 - x^8} + \cdots + \underbrace{x^6 - x^6} +$

$\underbrace{+x^5 - x^5} + x^5 + \underbrace{x^4 - x^4} + \cdots + \underbrace{x - x} + 1$ (now rearranging the terms)

$= x^{10} + x^9 + x^8 - x^9 - x^8 - x^7 + x^7 + x^6 + x^5 - x^6 - x^5 - x^4$

$+ x^5 + x^4 + x^3 - x^3 - x^2 - x + x^2 + x + 1$

$$= x^8(x^2 + x + 1) - x^7(x^2 + x + 1) + x^5(x^2 + x + 1) - x^4(x^2 + x + 1)$$
$$+ x^3(x^2 + x + 1) - x(x^2 + x + 1) + x^2 + x + 1$$
$$= (x^2 + x + 1)(x^8 - x^7 + x^5 - x^4 + x^3 - x + 1)$$

Python easily comes up with this factorisation:

```
[11]: factor(x**10 + x**5 + 1)
```

$[11]:$ $\left(x^2 + x + 1\right)\left(x^8 - x^7 + x^5 - x^4 + x^3 - x + 1\right)$

Exercise 6.1 *Factorise the polynomial* $(1 + x)^{30} + (1 - x)^{30}$.

Solution

The only challenge is to translate the expression correctly into Python.

```
[12]: factor((1 + x)**30 + (1 - x)**30)
```

$[12]:$

$$2\left(x^2 + 1\right)\left(x^4 + 14x^2 + 1\right)\left(x^8 + 44x^6 + 166x^4 + 44x^2 + 1\right)$$
$$\left(x^{16} + 376x^{14} + 4380x^{12} + 15944x^{10} + 24134x^8 + 15944x^6 + 4380x^4 + 376x^2 + 1\right)$$

Exercise 6.2 *Prove that the product of four consecutive numbers plus one is always a square number.*

Solution

Suppose the number is n. Then we are looking at $n(n + 1)(n + 2)(n + 3) + 1$. We ask Python to factorise this expression

```
[13]: n = symbols('n')
```

```
[14]: factor(n * (n + 1) * (n + 2) * (n + 3) + 1)
```

$[14]:$ $\left(n^2 + 3n + 1\right)^2$

The result clearly shows this expression is a square number. The proof is complete!

Exercise 6.3 *Using symbolic Python show that:*

$$\frac{1}{1 + \cfrac{1}{1 + \cfrac{1}{1 + \frac{1}{1+x}}}} = \frac{3 + 2x}{5 + 3x}.$$

Solution

If we define $f(x) = \frac{1}{1+x}$ then $f(f(x)) = \frac{1}{1+\frac{1}{1+x}}$ and $f(f(f(x))) = \frac{1}{1+\frac{1}{1+\frac{1}{1+x}}}$. This shows a way to capture the left-hand side of the above equality without going through the pain of typing it.

```
[15]: def f(x):
          return 1/(1+x)

      x = symbols('x')

      left_side = f(f(f(f(x))))
```

```
[16]: left_side
```

$$[16]: \quad \frac{1}{1 + \cfrac{1}{1 + \cfrac{1}{1 + \frac{1}{x+1}}}}$$

Next, in order to work with this expression, we import the method `simplify` from the `sympy` library. This function will try to, well, simplify the expressions as much as possible.

```
[17]: from sympy import simplify

      simplify(left_side)
```

$$[17]: \quad \frac{2x + 3}{3x + 5}$$

```
[18]: simplify(left_side) == (2*x +3)/(3*x +5)
```

```
[18]: True
```

Exercise 6.4 *Recall that if one wants to prove, by mathematical induction, that a statement P(n) is valid for all natural numbers n, one needs first to check P(1) is valid, and then assuming P(k) is correct, prove that P(k + 1) is also valid.*

Using symbolic Python and mathematical induction, prove the following identities:

$$1 + 2 + \cdots + n = \frac{n(n+1)}{2}$$

$$1^2 + 2^2 + \cdots + n^2 = \frac{n(n+1)(2n+1)}{6}.$$

Solution

We prove the first identity. Define a function as follows:

```
[19]: def g(n):
          return n * (n + 1) / 2
```

Clearly the identity is valid for $n = 1$

```
[20]: g(1)
```

```
[20]: 1.0
```

Now we suppose it is correct for k and check that it is also valid for $k + 1$. Thus we assume

$$1 + 2 + \cdots + k = \frac{k(k+1)}{2} = g(k)$$

and we need to show that

$$1 + 2 + \cdots + k + k + 1 = g(k) + k + 1 = g(k+1).$$

We ask symbolic Python for help, checking for equality of these two expression

```
[21]: import sympy

k = sympy.Symbol('k')
(g(k) + k + 1).equals(g(k + 1))
```

```
[21]: True
```

Thus we have proved, by mathematical induction, that the first identity is valid. Here is the code for the second identity.

```
[22]: def f(n):
          return n * (n + 1) * (2 * n + 1) / 6

f(1)
```

```
[22]: 1.0
```

```
[23]: import sympy

      n = sympy.Symbol('n')
      (f(n) + (n + 1)**2).equals(f(n + 1))
```

```
[23]: True
```

6.2 Graphics in sympy

It is always very helpful to present the behaviour of a function, an equation or a collection of data by plotting their graphs and visualising them. The dedicated library to handle graphics in Python is the library matplotlib, which is a very powerful library to plot the graph of data, as we will see in Chapter 7.

The sympy library comes with its own graphical methods as well. The sympy library's graphics allow us to plot the graphs of equations with ease. Under the hood, sympy is using the library matplotlib to generate the graphs.

For functions with one variable sympy offers a variety of commands to handle the plotting as mathematical equations can come in different forms. The following table shows which command is suitable for different formats of equations.

Two-dimensional mathematical function	Concrete example of 2-dimensional function	sympy method
$y = f(x)$	$y = \sin(x)/x$	plot
$x = f(t), y = g(t)$	$x = \sin(3t), y\cos(4t)$	plot_parametric
$f(x, y) = 0$	$x^4 - (x^2 - y^2) = 0$	plot_implicit, Eq
$f(x, y) > 0$	$x^4 + (x - 2y^2) > 0$	plot_implicit

Here we will showcase the graphical abilities of sympy with various interesting examples. We start by defining the function $f(x) = \sin(x)/x$ and plot its graph between -10π and 10π.

```
[24]: from sympy import sin, symbols, pi
      from sympy.plotting import plot

      def f(x):
          return sin(x) / x

      x = symbols('x')
```

[25]: `f(x)`

[25]: $\dfrac{\sin(x)}{x}$

[26]: `f(x**2)`

[26]: $\dfrac{\sin(x^2)}{x^2}$

[27]: `plot(f(x));`

[27]:
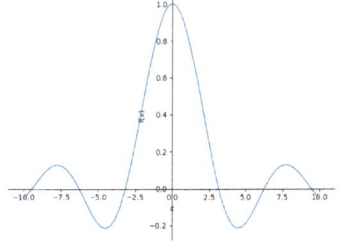

If we don't determine the range, as in the above example, Python uses its own default range (between -10, and 10.

[28]: `sympy.plotting.plot(f(x), (x, -10 * sympy.pi, 10 * sympy.`
       `↪pi));`

[28]:
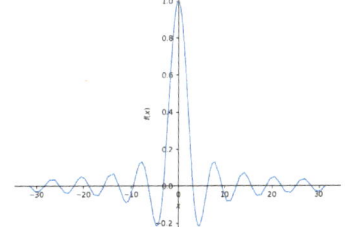

The `plot` method in `sympy.plotting` is capable of plotting a sequence of functions.

[29]: ```
plot(f(x), -f(x) + 2 , (x, -10 * pi, 10 * pi));
```

[29]:

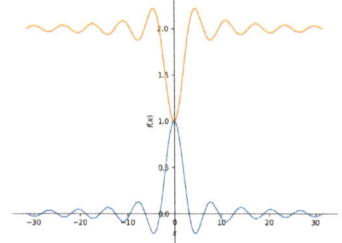

[30]: ```
s=[f(i * x) for i in range(1, 5)]
```

[31]: ```
s
```

[31]: `[sin(x)/x, sin(2*x)/(2*x), sin(3*x)/(3*x), sin(4*x)/(4*x)]`

Recall that *s will unpack the list into a sequence and thus we can get the plots of all the functions in the list s in one go.

[32]: ```
plot(*s);
```

[32]:

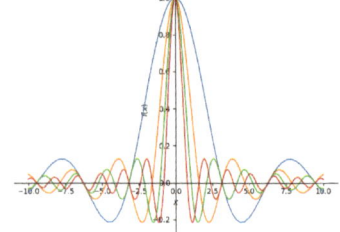

Exercise 6.5 *Define the function* $f(x) = ||x| - 1|$ *and plot* $f(x)$, $f(f(x))$ *and* $f(f(f(x)))$. *The absolute value function* $||$ *is defined as* **Abs** *in Python's* sympy *module.*

Solution

Let us understand this function. The absolute value of x is defined to be $-x$ if x is negative and x otherwise. We get:

$$f(x) = \begin{cases} x - 1 \text{ if } x \geq 1 \\ -x + 1 \text{ if } 0 \leq x < 1 \\ x + 1 \text{ if } -1 \leq x < 0 \\ -x - 1 \text{ if } x \leq -1. \end{cases}$$

The reader might imagine that for $f(f(x))$ one should consider several other cases. Once we have defined $f(x)$, we can ask Python to plot them for us.

```
[33]:  from sympy import sin, symbols, Abs
       from sympy.plotting import plot

       def f(x):
           return Abs(Abs(x) - 1)

       x = symbols('x')
       plot(f(x))
       plot(f(f(x)))
       plot(f(f(f(x))));
```

[33]:

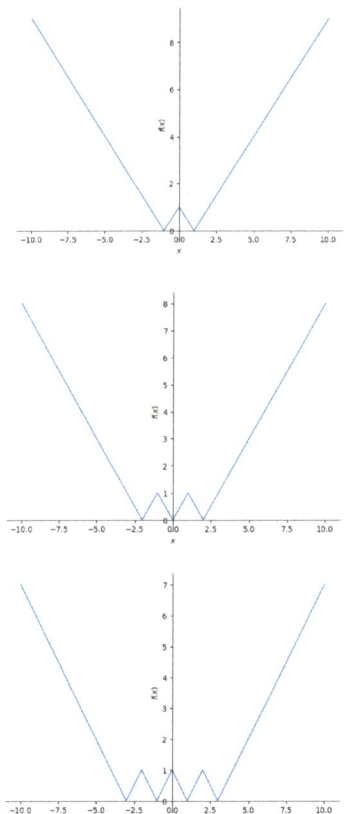

Next we plot the graph of the parametric equations defined by $x = \sin(3t)$, $y = \cos(4t)$.

```
[34]: from sympy import symbols, cos, sin
      from sympy import plot_parametric

      t = symbols('t')

      plot_parametric((sin(3 * t), cos(4 * t)), (t, -5, 5));
```

[34]: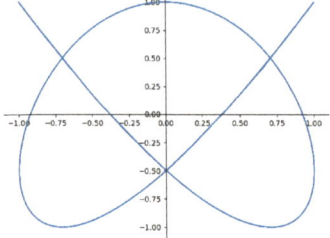

Exercise 6.6 *Draw the butterfly curve, discovered by Temple H. Fay, given by*

$$x(t) = \sin(t)\left(e^{\cos(t)} - 2\cos(4t) - \sin^5(t/12)\right),$$
$$y(t) = \cos(t)\left(e^{\cos(t)} - 2\cos(4t) - \sin^5(t/12)\right).$$

Solution

Clearly, the coordinates (x, y) in the equation depend on a variable t. Thus we need to use plot_parametric to handle this equation. We define functions $x(t)$ and $y(t)$, and once we have these ready we pass them into plot_parametric; sympy can handle the rest.

```
[35]: from sympy import symbols, cos, sin, exp
      from sympy import plot_parametric

      t = symbols('t')

      def x(t):
          return sin(t) * (exp(cos(t)) - 2 * cos(4 * t) - sin(t/
       ↪12)**5)

      def y(t):
          return cos(t) * (exp(cos(t)) - 2 * cos(4*t) - sin(t/
       ↪12)**5)

      plot_parametric(x(t), y(t), (t, -50, 50));
```

[35]: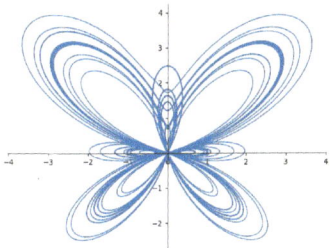

Next we find the graph of all the points (x, y) which satisfy the equation $x^4 - (x^2 - y^2) = 0$. For such equations, one uses plot_implicit as follows:

[36]:
```python
from sympy import symbols, Eq
from sympy import plot_implicit

x, y = symbols('x y')

plot_implicit(Eq(x**4 - (x**2 - y**2), 0), (x, -1.5, 1.5),↵
↪(y, -1.5, 1.5));
```

[36]: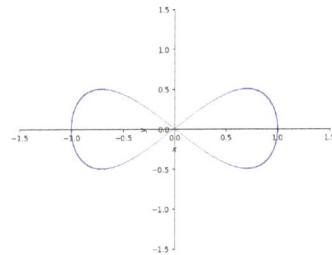

Note the use of Eq to define the equation $x^4 - (x^2 - y^2) = 0$.

Exercise 6.7 *Consider the equation of the so-called bouncing wagon*

$$2y^3 + y^2 - y^5 = x^4 - 2x^3 + x^2.$$

Plot it and notice the interesting graph!

Solution

Can you see the bouncing wagon in the picture?

[37]:
```python
from sympy import symbols, Eq, sin
from sympy import plot_implicit

x, y = symbols('x y')
```

```
plot_implicit(Eq(2*y**3 + y**2 - y**5, x**4 - 2*x**3 + x**2),↵
 ↵(x, -5, 5), (y, -5, 5));
```

[37]:

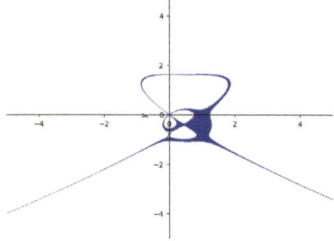

We can plot several graphs on the same x-y-plane. For example, we would like to know how many pairs of real numbers (x, y) satisfy the system of equations

$$2 - x^3 = y$$
$$2 - y^3 = x + \sin(y).$$

By plotting the graphs of these two equations, we should be able to deduce the number of solutions.

[38]:
```
p1 = plot_implicit(Eq(2 - x**3, y), (x, -5, 5), (y, -5, 5),↵
 ↵line_color='blue', show=False);
p2 = plot_implicit(Eq(2 - y**3, x + sin(y)), (x, -5, 5), (y,↵
 ↵-5, 5), line_color='green',show=False);
p1.append(p2[0])

p1.show()
```

[38]:

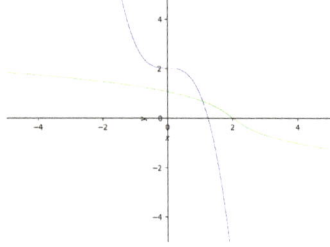

Note that here $p1$ and $p2$ are two objects. With show=False, we initially don't plot the graph. We create the object and then add the second one to the first one and then plot this new object.

To obtain the region of all points (x, y) which satisfy the inequality $x^4 + (x - 2y^2) > 0$, we can use plot_implicit.

[39]:
```
from sympy import plot_implicit

plot_implicit(x**4 + (x - 2*y**2) > 0, (x, -2, 2), (y, -2,␣
→2));
```

[39]: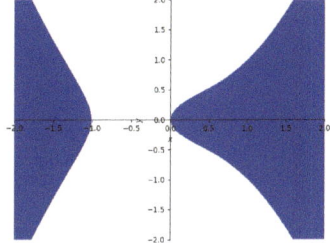

6.3 Three-Dimensional Graphs

In a similar manner, one can define functions of several variables. The chart below
shows what command one needs to choose to plot functions defined in different
formats.

Three-dimensional mathematical function	Concrete example of 3-dimensional function	sympy method
$y = f(x, y)$	$y = \sin(z^2 + y^2)e^{-x^2}$	plot3d
$\begin{cases} x = f(t), \\ y = g(t), \\ z = h(t) \end{cases}$	$\begin{cases} x = \sin(3t), \\ y = \cos(4t), \\ z = \sin(5t) \end{cases}$	plot3d_parametric_line
$\begin{cases} x = f(t, u), \\ y = g(t, u), \\ z = h(t, u) \end{cases}$	$\begin{cases} x = \sin(3t), \\ y = \cos(4t), \\ z = \sin(5t) \end{cases}$	plot3d_parametric_surface

Here is a simple example defining $f(x, y) = \sqrt{x^2 + y^2}$

[40]:
```
from sympy import sqrt

def f(x, y):
    return sqrt(x**2 - y**2)

f(x, y)
```

[40]: $\sqrt{x^2 - y^2}$

Once the function is defined, it is very easy to plot its graph via the sympy library.

[41]:
```
from sympy.plotting import plot3d

plot3d(f(x, y));
```

[41]:

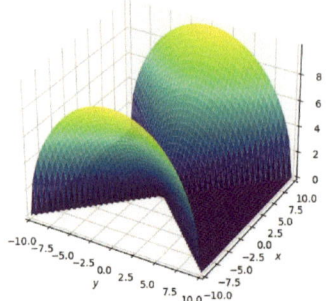

[42]:
```
plot3d(f(x, y), (x, -100, 100), (y, -100, 100));
```

[42]:

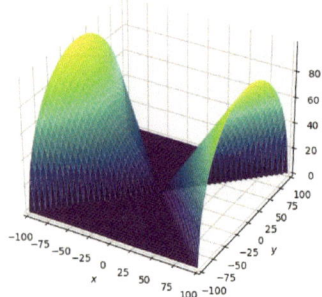

Note that, for two-dimensional equations, such as $f(x) = x + \sin(x)$, one uses plot whereas for three-dimensional equations, such as $\sqrt{x^2 + y^2}$, one uses plot3d.

Exercise 6.8 *Plot the graph of the "cowboy hat" equation*

$$\sin(x^2 + y^2)e^{-x^2} + \cos(x^2 + y^2)$$

as both x and y range from −2 to 2.

Solution

We first translate the formula into Python and use plot3d within sympy to create the graph.

```
[43]:  from sympy import sin, cos, exp
       from sympy.plotting import plot3d

       x, y = symbols('x y')

       plot3d(sin(x**2 + y**2) * exp(-x**2) + cos(x**2 + y**2));
```

[43]:

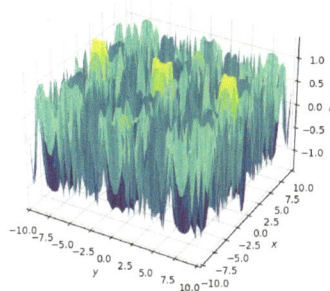

This is not the picture we were expecting. Restricting the range as explicitly mentioned in the exercise we get:

```
[44]:  plot3d(sin(x**2+y**2)*exp(-x**2) + cos(x**2+y**2),(x, -2,
       ↪2),(y, -2, 2));
```

[44]:

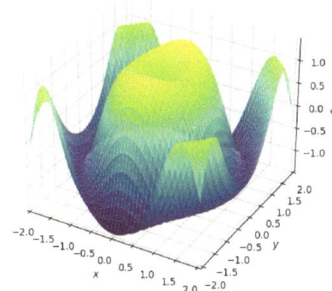

Exercise 6.9 *Define the function* $f(x, y) = ||x| - |y||$ *and plot its graph for* $-10 \leq x, y \leq 10$.

Solution

For plotting a function with two variables we can use the function plot3d of the sympy.plotting module.

```
[45]: from sympy import symbols, Abs
      from sympy.plotting import plot3d

      def f(x, y):
          return Abs(Abs(x) - Abs(y))

      x, y = symbols('x y')

      plot3d(f(x, y), (x, -10, 10), (y, -10, 10));
```

[45]:

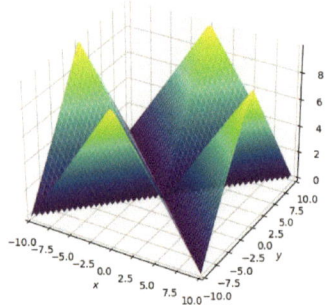

Exercise 6.10 *Plot the graph of $f(x, y) = xy \sin(x^2) \cos(y^2)$ when $-2\pi \le x \le 0$ and $-2\pi \le y \le 0$.*

Solution

In the previous exercise, we first defined a function $f(x, y)$, and then used the function within plot3d. Here we directly use the equation inside plot3d.

```
[46]: from sympy import symbols, sin, cos, pi
      from sympy.plotting import plot3d

      x, y = symbols('x y')

      plot3d(x * y * sin(x**2) * cos(y**2), (x, -2*pi, 0), (y,↵
      ↪-2*pi, 0));
```

[46]:

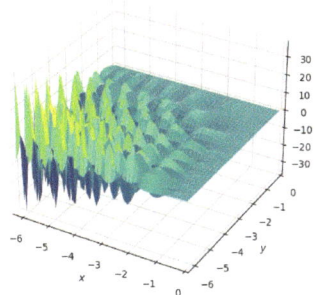

Next we look at a 3-dimensional parametric plot. For $-\pi \le t \le \pi$, plot the graph of

$$x = \sin(3t),$$
$$y = \cos(4t),$$
$$z = \sin(5t).$$

[47]:
```
from sympy import symbols, cos, sin, pi
from sympy.plotting import plot3d_parametric_line

t = symbols('t')
plot3d_parametric_line(sin(3*t), cos(4*t), sin(5*t), (t, -pi,␣
 ↪pi));
```

[47]:

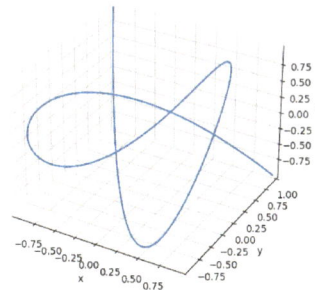

[48]:
```
t = symbols('t')
plot3d_parametric_line(sin(t), cos(t), t, (t, -10*pi,␣
 ↪10*pi));
```

[48]:

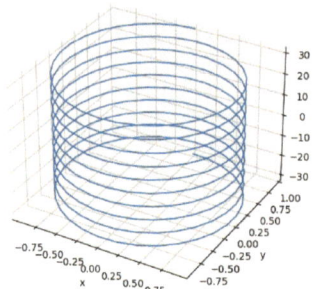

6.4 Calculus with sympy

6.4.1 Solving equations

Solving equations and finding roots for different types of equations and relations are one of the main endeavours of mathematics. For polynomials with one variable, i.e., of the form $a_n x^n + a_{n-1} x^{n-1} + \cdots + a_1 x + a_0$, it has been proved that there is no formula for finding the roots when $n \geq 5$ (in fact, when $n = 3$ or 4, the formulas are not that pretty!). This forces us to find numerical ways to estimate the roots of equations. However, if it is possible, we might be able to find the exact solutions to an equation using sympy.

As a first example, we use sympy to find the solutions of the equation $x^2 - 3x - 10 = 0$.

```
[49]: from sympy import solve, solveset

      solve(x**2 - 3*x - 10, x)
```

[49]: [-2, 5]

```
[50]: solveset(x**2 - 3*x - 10, x)
```

[50]: $\{-2, 5\}$

Here we use the function solve, which is available in the sympy library. The second argument in the function tells Python x is the variable of the equation. As the result shows, the command solveset returns a set containing the solutions.

Plotting the graph of the polynomial $x^2 - 3x - 10$ confirms that -2 and 5 are the roots of the equation $x^2 - 3x - 10 = 0$.

```
[51]: from sympy.plotting import plot

      plot(x**2 - 3*x - 10);
```

[51]:

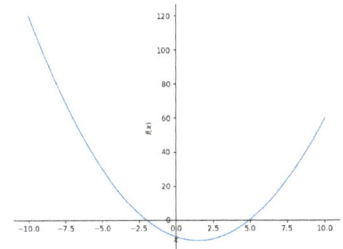

Next we try to solve the equation $x^4 - 3x^3 + 2x + 10 = 0$

```
[52]: from sympy.plotting import plot

      plot(x**4 - 3*x**3 + 2*x + 10);
```

[52]:

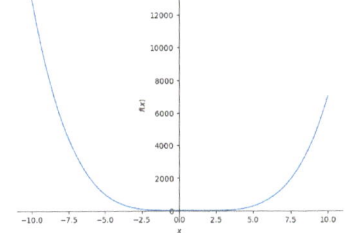

```
[53]: from sympy import solveset, simplify

      solveset(x**4 - 3*x**3 + 2*x + 10, x)
```

[53]:

$$\frac{3}{4} - \frac{i\sqrt{-\frac{9}{2} + \frac{23}{3\sqrt[3]{\frac{47}{8} + \frac{\sqrt{113079}i}{72}}} - \frac{11}{4\sqrt{\frac{9}{4} + \frac{23}{3\sqrt[3]{\frac{47}{8} + \frac{\sqrt{113079}i}{72}}} + 2\sqrt[3]{\frac{47}{8} + \frac{\sqrt{113079}i}{72}}}} + 2\sqrt[3]{\frac{47}{8} + \frac{\sqrt{113079}i}{72}}}}{2} +$$

$$\frac{\sqrt{\frac{9}{4} + \frac{23}{3\sqrt[3]{\frac{47}{8} + \frac{\sqrt{113079}i}{72}}} + 2\sqrt[3]{\frac{47}{8} + \frac{\sqrt{113079}i}{72}}}}{2},$$

$$\frac{3}{4} - \frac{i\sqrt{-\frac{9}{2}+\frac{23}{3\sqrt[3]{\frac{47}{8}+\frac{\sqrt{113079}i}{72}}}+\frac{11}{4\sqrt{\frac{9}{4}+\frac{23}{3\sqrt[3]{\frac{47}{8}+\frac{\sqrt{113079}i}{72}}}+2\sqrt[3]{\frac{47}{8}+\frac{\sqrt{113079}i}{72}}}}+2\sqrt[3]{\frac{47}{8}+\frac{\sqrt{113079}i}{72}}}}{2} - $$

$$\frac{\sqrt{\frac{9}{4}+\frac{23}{3\sqrt[3]{\frac{47}{8}+\frac{\sqrt{113079}i}{72}}}+2\sqrt[3]{\frac{47}{8}+\frac{\sqrt{113079}i}{72}}}}{2},$$

$$\frac{3}{4} - \frac{\sqrt{\frac{9}{4}+\frac{23}{3\sqrt[3]{\frac{47}{8}+\frac{\sqrt{113079}i}{72}}}+2\sqrt[3]{\frac{47}{8}+\frac{\sqrt{113079}i}{72}}}}{2} + $$

$$\frac{i\sqrt{-\frac{9}{2}+\frac{23}{3\sqrt[3]{\frac{47}{8}+\frac{\sqrt{113079}i}{72}}}+\frac{11}{4\sqrt{\frac{9}{4}+\frac{23}{3\sqrt[3]{\frac{47}{8}+\frac{\sqrt{113079}i}{72}}}+2\sqrt[3]{\frac{47}{8}+\frac{\sqrt{113079}i}{72}}}}+2\sqrt[3]{\frac{47}{8}+\frac{\sqrt{113079}i}{72}}}}{2},$$

$$\frac{3}{4} + \frac{\sqrt{\frac{9}{4}+\frac{23}{3\sqrt[3]{\frac{47}{8}+\frac{\sqrt{113079}i}{72}}}+2\sqrt[3]{\frac{47}{8}+\frac{\sqrt{113079}i}{72}}}}{2} + $$

$$\frac{i\sqrt{-\frac{9}{2}+\frac{23}{3\sqrt[3]{\frac{47}{8}+\frac{\sqrt{113079}i}{72}}}-\frac{11}{4\sqrt{\frac{9}{4}+\frac{23}{3\sqrt[3]{\frac{47}{8}+\frac{\sqrt{113079}i}{72}}}+2\sqrt[3]{\frac{47}{8}+\frac{\sqrt{113079}i}{72}}}}+2\sqrt[3]{\frac{47}{8}+\frac{\sqrt{113079}i}{72}}}}{2}$$

The reader can try for themselves to see solve cannot give any solutions to the equation $\sin(x) = x - 1$, although there is one real solution to this equation, as the following graph shows.

[54]:
```
from sympy import sin
from sympy.plotting import plot

plot(sin(x) - x + 1);
```

[54]:

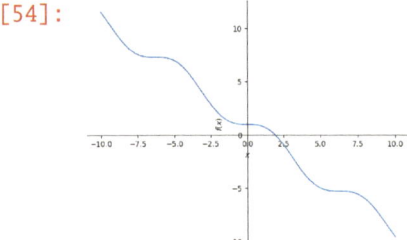

The graph shows there is solution around 2. We could write a simple program to estimate the root.

```
[55]: from math import sin

      x = 1.5
      while sin(x) - x + 1 > 0.00001:
          x += 0.001

      print(f'{x} and {sin(x) - x + 1}')
```

```
[55]: 1.934999999999952 and -0.0005922867585996805
```

6.4.2 Limits, derivation and integration

Two important machineries in calculus are differentiation and integration, and both use the concept of limit. We assume the reader is familiar with calculus. In the problems below we showcase some of sympy's abilities in this area.

We start with the limit of the function

$$\lim_{x \to 0} \frac{\cos(x) - 1}{\sin(x)}.$$

sympy provides the command limit for exploring the limit of a function.

```
[56]: from sympy import sin, cos, pi
      from sympy import limit

      x = symbols('x')

      limit((cos(x)-1)/sin(x), x, 0)
```

```
[56]: 0
```

The graph of the function confirms that, when x tends to 0, the limit of this function is indeed 0.

```
[57]: from sympy import sin, cos
      from sympy.plotting import plot

      plot((cos(x)-1)/sin(x), (x, -3.1, 3.1));
```

[57]: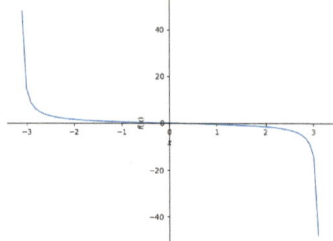

We can also explicitly ask sympy to calculate the limit when the variable tends from the right or left to a given value.

[58]: `limit((cos(x)-1)/sin(x), x, pi, '-')`

[58]: $-\infty$

[59]: `limit((cos(x)-1)/sin(x), x, pi, '+')`

[59]: ∞

Exercise 6.11 *Investigate*

$$\lim_{n \to 0} \frac{\sin(x)}{x} = 1.$$

Solution

This limit tells us that, if the value of x is very small, one can essentially replace $\sin(x)$ with x. We first plot the graph of the equation $\sin(x)/x$ as well as the constant line 1.

[60]:
```
from sympy import sin, cos, pi
from sympy.plotting import plot

plot(sin(x)/x, 1,  (x, -pi, pi));
```

[60]: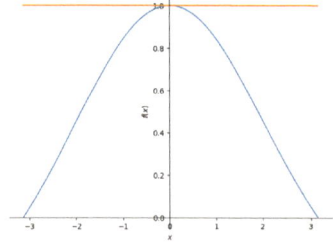

This clearly shows that the constant line 1 touches (and is tangent to) the graph of $\sin(x)/x$ at $x = 0$. We can confirm this with the sympy command `limit`.

```
[61]: limit(sin(x)/x, x, 0)
```

[61]: 1

Next we look at derivatives and integrals. The following table gives an overview of
how sympy handles these operations.

Examples of functions	Commands to use
$\partial f/\partial x$	diff(f,x)
$\partial^2 f/\partial x\partial x$	diff(f,x,x) or diff(f,x,2)
$\partial^2 f/\partial x\partial y$	diff(f,x,y)
$\int f(x)dx$	Integrate(f(x),x)
$\int_a^b f(x)dx$	Integrate(f(x),x)
$\int_c^d \int_a^b f(x,y)dxdy$	Integrate(f(x),(x, a,b), (y, c,d))

One can use oo and -oo for ∞ and $-\infty$ in the calculations above.

As a first example, we calculate

$$\int_0^\infty e^{-x}dx.$$

```
[62]: from sympy import exp, integrate, symbols, oo
      from sympy.plotting import plot

      x= symbols('x')

      integrate(exp(-x), (x, 0, oo))
```

[62]: 1

This means the area under the curve e^{-x} from 0 to ∞ is exactly 1.

```
[63]: plot(exp(-x), (x, 0, 10))
```

[63]:

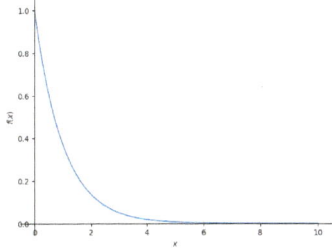

Exercise 6.12 *Evaluate the following:*

$$\frac{\partial f}{\partial x}, \text{ when } f = \sin(x)/x,$$

$$\frac{\partial^2 f}{\partial x^2}, \text{ when } f = \sin(x)/x,$$

$$\frac{\partial^3 f}{\partial x^2 \partial y}, \text{ when } f = e^{xy},$$

$$\int \left(\cos(x)/x - \sin(x)/x^2 \right) dx$$

$$\int_{-1}^{1} \int_{-1}^{1} \cos(x^2 + y^2 + xy) dx dy.$$

Solution

We start from the top and work our way down. We need the methods `integrate` and `diff` from the `sympy` library in order to handle the integration and derivation.

```
[64]: from sympy import exp, integrate, diff, symbols, oo
      from sympy.plotting import plot
```

```
[65]: diff(sin(x)/x)
```

$$[65]: \quad \frac{\cos(x)}{x} - \frac{\sin(x)}{x^2}$$

```
[66]: diff(sin(x)/x, x, 2)
```

$$[66]: \quad \frac{-\sin(x) - \frac{2\cos(x)}{x} + \frac{2\sin(x)}{x^2}}{x}$$

```
[67]: diff(sin(x)/x, x, x)
```

$$[67]: \quad \frac{-\sin(x) - \frac{2\cos(x)}{x} + \frac{2\sin(x)}{x^2}}{x}$$

To obtain an expression in the form p/q, one can use the command `cancel` within the library sympy. Recall that _ refers to the last output.

```
[68]: from sympy import cancel

      cancel(_)
```

[68]: $\dfrac{-x^2 \sin(x) - 2x \cos(x) + 2 \sin(x)}{x^3}$

[69]: `diff(exp(x * y), x, x, y)`

[69]: $y(xy + 2) e^{xy}$

[70]: `integrate(cos(x)/x - sin(x)/x**2, x)`

[70]: $\dfrac{\sin(x)}{x}$

The last integration takes some time to be calculated. This is because sympy will try to give a precise form of the answer. We will see that a numerical approach will produce the result much faster.

[71]: `integrate(cos(x**2 + y**2 + x*y), (x, -1, 1), (y, -1, 1))`

[71]:

$$\frac{\sqrt{2}\sqrt{\pi}}{2} \times \Bigg(\int_{-1}^{1} \sin\left(\frac{3y^2}{4}\right) S\left(\frac{\sqrt{2}y}{2\sqrt{\pi}} - \frac{\sqrt{2}}{\sqrt{\pi}}\right) dy +$$

$$\int_{-1}^{1} \left(-\sin\left(\frac{3y^2}{4}\right) S\left(\frac{\sqrt{2}y}{2\sqrt{\pi}} + \frac{\sqrt{2}}{\sqrt{\pi}}\right)\right) dy +$$

$$\int_{-1}^{1} \cos\left(\frac{3y^2}{4}\right) C\left(\frac{\sqrt{2}y}{2\sqrt{\pi}} + \frac{\sqrt{2}}{\sqrt{\pi}}\right) dy +$$

$$\int_{-1}^{1} \left(-\cos\left(\frac{3y^2}{4}\right) C\left(\frac{\sqrt{2}y}{2\sqrt{\pi}} - \frac{\sqrt{2}}{\sqrt{\pi}}\right)\right) dy \Bigg)$$

[72]: `round(_)`

[72]: 3

Exercise 6.13 *Consider the two surfaces*

$$q(x, y) = \cos(x^2 + y^2) \exp(-x^2),$$
$$w(x, y) = 3 - x^2 - y^2.$$

Plot these functions.

Solution

We have seen examples of this type. The only thing to notice is we can use `plot3d` to plot a sequence of functions, as the code shows.

```
[73]: from sympy import symbols, sin, cos, pi, exp
      from sympy.plotting import plot3d

      x, y = symbols('x y')

      plot3d(cos(x**2 + y**2) * exp(-x**2), 3 - x**2 - y**2, (x,␣
          ↪-3, 3), (y, -3, 3));
```

[73]:

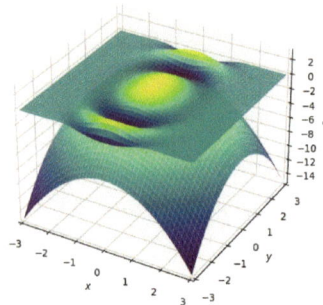

Exercise 6.14 Let $A = (a_{ij})$ denote the 4×4 matrix with

$$a_{ij} = x^i + x^j + x^{ij}.$$

Symbolically compute the determinant of A? If $x = \sqrt{2}$, what is the inverse of the matrix A?

Solution

Note that the entries of the matrix contain the symbol x. Thus sympy is the library to use here to define the function. We can pass the symbol x inside the function m. This function puts together the matrix. It consists of a nested loop, the first loop taking care of the rows and the other loop putting together the entries of each row.

```
[74]: from sympy import symbols, det, Matrix, sqrt, factor

      def m(x):
          A = []
          for i in range(1, 5):
              row = []
              for j in range(1, 5):
                  row += [x**i + x**j + x**(i*j)]
              A += [row]
          return Matrix(A)
```

```
x = symbols('x')
A = m(x)

A
```

[74]:
$$\begin{bmatrix} 3x & 2x^2+x & 2x^3+x & 2x^4+x \\ 2x^2+x & x^4+2x^2 & x^6+x^3+x^2 & x^8+x^4+x^2 \\ 2x^3+x & x^6+x^3+x^2 & x^9+2x^3 & x^{12}+x^4+x^3 \\ 2x^4+x & x^8+x^4+x^2 & x^{12}+x^4+x^3 & x^{16}+2x^4 \end{bmatrix}$$

[75]: `A.det()`

[75]: $3x^{30} - 10x^{29} + 6x^{28} + 11x^{27} - 8x^{26} - 10x^{25} - 2x^{24} + 21x^{23} - 3x^{22} - 11x^{21} - 6x^{20} + 6x^{19} + 10x^{18} - 4x^{17} - 2x^{16} - 7x^{15} + 6x^{14} + 3x^{13} - 4x^{12} + x^{11}$

[76]: `factor(A.det())`

[76]: $x^{11}(x-1)^6(x+1)^2\left(x^2+x+1\right)\left(3x^9-x^8+2x^5-x^2-x+1\right)$

For the final part of the exercise, to compute the inverse of the matrix when $x = \sqrt{2}$, we first pass $\sqrt{2}$ into the function m(x), and then use the method inv within this object to compute its inverse. The computation is impressive and should not be lost on the reader.

[77]: `mat_inv = m(sqrt(2)).inv()`

[78]: `mat_inv`

[78]:
$$\begin{bmatrix} \frac{1766118-1248857\sqrt{2}}{-5587390+3950878\sqrt{2}} & \frac{-480344+339683\sqrt{2}}{-5587390+3950878\sqrt{2}} & \frac{-730010+516175\sqrt{2}}{-11174780+7901756\sqrt{2}} & \frac{172154-121729\sqrt{2}}{-11174780+7901756\sqrt{2}} \\ \frac{-339683+240172\sqrt{2}}{-3950878+2793695\sqrt{2}} & \frac{-2165050+1531031\sqrt{2}}{-15803512+11174780\sqrt{2}} & \frac{181333-128264\sqrt{2}}{-15803512+11174780\sqrt{2}} & \frac{169553-119887\sqrt{2}}{-15803512+11174780\sqrt{2}} \\ \frac{-25415+17965\sqrt{2}}{-388988+275054\sqrt{2}} & \frac{8941-6304\sqrt{2}}{-777976+550108\sqrt{2}} & \frac{55676-39385\sqrt{2}}{-1555952+1100216\sqrt{2}} & \frac{-12698+8981\sqrt{2}}{-1555952+1100216\sqrt{2}} \\ \frac{2251-1604\sqrt{2}}{-146656+103738\sqrt{2}} & \frac{3175-2206\sqrt{2}}{-293312+207476\sqrt{2}} & \frac{-4816+3367\sqrt{2}}{-586624+414952\sqrt{2}} & \frac{-1786+1269\sqrt{2}}{-586624+414952\sqrt{2}} \end{bmatrix}$$

[79]: `mat = m(sqrt(2))`

[80]: `mat`

[80]:
$$\begin{bmatrix} 3\sqrt{2} & \sqrt{2}+4 & 5\sqrt{2} & \sqrt{2}+8 \\ \sqrt{2}+4 & 8 & 2\sqrt{2}+10 & 22 \\ 5\sqrt{2} & 2\sqrt{2}+10 & 20\sqrt{2} & 2\sqrt{2}+68 \\ \sqrt{2}+8 & 22 & 2\sqrt{2}+68 & 264 \end{bmatrix}$$

Next we multiply these two matrices; we are expecting the 4×4-identity matrix, however we obtain a massive expression. We will not try to display the output here

as it will extend beyond the margin (this is not Fermat's comment!). But we will ask
Python to simplify the output.

[81]: `mat * mat_inv;`

[82]: `simplify(_)`

[82]: $\begin{bmatrix} 1 & 0 & 0 & 0 \\ 0 & 1 & 0 & 0 \\ 0 & 0 & 1 & 0 \\ 0 & 0 & 0 & 1 \end{bmatrix}$

Problems

1. Plot the graph of

$$x(t) = 4\cos(-11t/4) + 7\cos(t),$$
$$y(t) = 4\sin(-11t/4) + 7\sin(t)$$

for $0 \le t \le 14\pi$.

2. Plot the graph of

$$x(t) = \cos(t) + 1/2\cos(7t) + 1/3\sin(17t)$$
$$y(t) = \sin(t) + 1/2\sin(7t) + 1/3\cos(17t)$$

for $0 \le t \le 14\pi$.

3. Consider the following function.

$$f(x) := \sum_{k=1}^{100} \left(\frac{\sin\left(2\pi k^2 x\right)}{4\pi^2 k^5} + \frac{x^2}{2k} \right)$$

This function was suggested by Sungkon Chang as an example of a function
that looks quite "innocent" but whose derivatives behave quite wildly. Plot $f(x)$,
$f'(x)$ and $f''(x)$ and observe this behaviour.

4. Consider the following functions of two variables

$$\mathbf{x}(u, v) = \sin(v)\cos(u),$$
$$\mathbf{y}(u, v) = \sin(v)\sin(u),$$
$$\mathbf{z}(u, v) = \cos(v).$$

Generate the surface $(\mathbf{x}, \mathbf{y}, \mathbf{z})$ when $0 \le u \le 3\pi/2$ and $0 \le v \le \pi$.

Now consider $x_1(u, v) = \frac{-3}{8} \cos(v) \sin(\frac{4u}{3})$, $y_1(u, v) = \frac{3}{8} \cos(\frac{4u}{3}) \cos(v)$ and $z_1(u, v) = \frac{\sin(v)}{2}$. Generate the surface

$$\left(\frac{d x_1}{dvdu}, \frac{d y_1}{dudv}, \frac{d z_1}{dv} \right)$$

when $0 \le u \le 3\pi/2$ and $0 \le v \le \pi$. Finally, superimpose these two images.

5. Consider the function $\cos(x^2)$. Calculate the area below this function and between 0 and where the function first hits the x-axis.

6. Write a function to check that, for any n, the following identity holds:

$$\det \begin{pmatrix} x & a_1 & a_2 & \cdots & a_n \\ a_1 & x & a_2 & \cdots & a_n \\ a_1 & a_2 & x & \cdots & a_n \\ \vdots & \vdots & \vdots & & \vdots \\ a_1 & a_2 & a_3 & \cdots & x \end{pmatrix} = (x + a_1 + \cdots + a_n)(x - a_1) \cdots (x - a_n)$$

7. Write a function to check that, for any n, the following identity holds:

$$\det \begin{pmatrix} 1 & 1 & 1 & \cdots & 1 & 1 \\ b_1 & a_1 & a_1 & \cdots & a_1 & a_1 \\ b_1 & b_2 & a_2 & \cdots & a_2 & a_2 \\ \vdots & \vdots & \vdots & & \vdots & \vdots \\ b_1 & b_2 & b_3 & \cdots & b_n & a_n \end{pmatrix} = (a_1 - b_1)(a_2 - b_2) \cdots (a_n - b_n)$$

8. The Hilbert matrix is a square matrix whose element in row i and column j is $\frac{1}{i+j-1}$. Construct the Hilbert matrix of order 6 (i.e. a 6×6-matrix) and compute its determinant and its inverse. Try this for 7×7 and 8×8 matrices. Construct a table which shows the determinant of Hilbert matrices of orders 1 through 10.

9. Construct a 10×10 upper triangular matrix A (see Wikipedia for the definition of such matrices) whose nonzero entries are random integers. Show that its determinant is equal to the product of the entries on its main diagonal. Now consider the transpose matrix A^t and show that its determinant $(\det(A^t))$ is the same as A. What is $\det(A + A^t)$?

10. Investigate how many solutions the equation $\sin(x^2) - \cos(x^3) = 0$ has for $0 \le x \le \pi$.

11. Plot the graphs of the functions $2 \exp(-x^2)$ and $\cos(\sin(x) + \cos(x))$ between $[-\pi, \pi]$. Investigate where they intersect.

12. Define the functions

$$f(t, a) = 2 + \frac{1}{2}\sin(at),$$

$$g(t, b, c) = \cos\left(t + \frac{\sin(bt)}{c}\right),$$

$$h(t, b, c) = \sin\left(t + \frac{\sin(bt)}{c}\right).$$

Generate the parametric graph

$$x(t) = f(t, 8)g(t, 16, 4)$$
$$y(t) = f(t, 8)h(t, 16, 4)$$

when $0 \le t \le 2\pi$. Also plot the graph

$$(f(t, 6)g(t, 18, 18), f(t, 6)h(t, 18, 18))$$

when $0 \le t \le 2\pi$.

13. Let $g(x) = \sin(x) + \cos(x)$. Plot graphs of $g(g(g(g(x))))$ and $g(g(g(g(g(g(x))))))$ for x lying between 0 and π. There are four points where the graphs cross. Find, numerically, their (x, y) coordinates.

14. Let A denote the 3×3 matrix

$$\begin{pmatrix} x + y & x^2 + y & x^3 + y \\ x + y^2 & x^2 + y^2 & x^3 + y^2 \\ x^2 + y^3 & x^2 + y^3 & x^3 + y^3 \end{pmatrix}$$

(yes, the fact that the bottom left-hand entry does not fit the pattern of the rest of the entries is intended!) and let B denote its inverse. Show that the sum of the entries in the first row of B is 0. What is the sum of the entries in each of the second and third rows? Also find the sum of the entries in each column of B.

15. Investigate and determine the values of x between $0 \le x \le 2\pi$ such that we have the inequality

$$2\cos(x) \le \sqrt{1 + \sin(2x)} - \sqrt{1 - \sin(2x)} \le \sqrt{2}.$$

Chapter 7
The numpy Library

7.1 numpy, Numerical Python

In this chapter we will look at one of the most powerful and most used libraries of Python, that is, numpy, which is used working with data. The library numpy allows us to handle large lists of values; here a list is called an array. It provides a powerful way to do arithmetic on arrays. We can use arrays to handle large collections of inputs, and to handle matrices and thus linear algebra, one of the branches of mathematics that has become vital in all aspects of computer science.

As is customary, we import numpy as the alias np.

```
[1]: import numpy as np
```

Next, we introduce the arrays in numpy.

```
[2]: x = np.array([1, -2, 3.5])
```

```
[3]: x
```

```
[3]: array([ 1. , -2. ,  3.5])
```

```
[4]: list(x)
```

```
[4]: [1.0, -2.0, 3.5]
```

Notice how we can convert sequences and lists into an array.

```
[5]: y = np.array(range(1, 18, 2))
```

```
[6]: y
```

© The Author(s), under exclusive license to Springer Nature Switzerland AG 2023
R. Hazrat, *A Course in Python*, Springer Undergraduate Mathematics Series,
https://doi.org/10.1007/978-3-031-49780-3_7

```
[6]: array([ 1,  3,  5,  7,  9, 11, 13, 15, 17])
```

```
[7]: z = np.array([i % 2 for i in range(10)])
```

```
[8]: z
```

```
[8]: array([0, 1, 0, 1, 0, 1, 0, 1, 0, 1])
```

We can create multidimensional arrays, which we can also view as matrices.

```
[9]: t = np.array([[1, 2, 3], [4, 5, 6]])
```

```
[10]: t
```

```
[10]: array([[1, 2, 3],
             [4, 5, 6]])
```

What we have done so far is to convert lists or tuples into numpy arrays. There are functions within numpy which allow us to directly create arrays.

```
[11]: z = np.zeros((3, 6))
```

```
[12]: z
```

```
[12]: array([[0., 0., 0., 0., 0., 0.],
             [0., 0., 0., 0., 0., 0.],
             [0., 0., 0., 0., 0., 0.]])
```

```
[13]: z.shape
```

```
[13]: (3, 6)
```

7.1.1 Calculus on arrays

In principal all operations on arrays are done "component-wise". If we add two arrays, we add the entries of the arrays together. If we apply a function to an array, the function will be applied to each entry (similar to mapping a function into a list in functional programming).

```
[14]: ar = np.array([1, 2, 3])
```

```
[15]: ar + ar
```

```
[15]: array([2, 4, 6])
```

```
[16]: ra = ar * 2
```

```
[17]: ra
```

```
[17]: array([2, 4, 6])
```

```
[18]: ra - (ar + ar) == np.zeros(3)
```

```
[18]: array([ True,   True,   True])
```

Notice in the line above the comparison is done for each entry and thus we get three boolean True values.

7.1.2 Generating arrays

Besides creating arrays in numpy from lists or tuples as we have done so far, there are several functions within numpy which we can use to comfortably generate arrays. The two main ones are arange and linspace. The following examples show their scopes.

```
[19]: np.arange(5, 10)
```

```
[19]: array([5, 6, 7, 8, 9])
```

```
[20]: np.arange(5, 10, 0.3)
```

```
[20]: array([5., 5.3, 5.6, 5.9, 6.2, 6.5, 6.8, 7.1, 7.4, 7.7, 8.,
       8.3, 8.6, 8.9, 9.2, 9.5, 9.8])
```

Similar to range, arange(m, n, s) starts from m, and increases by s in each iteration *up* to but not including n. The command linspace(m, n, s) gives an array with s equally spaced entries, starting from m and ending with n; that is, the array starts with m and in each iteration $\frac{n-m}{s-1}$ will be added until we arrive at n.

```
[21]: np.linspace(1, 4, 6)
```

```
[21]: array([1., 1.6, 2.2, 2.8, 3.4, 4. ])
```

This command is quite handy when plotting graphs; we divide the area of interest using linspace, feed the corresponding values into the function and plot the result.

```
[22]: import matplotlib.pyplot as plt

      x = np.linspace(- 2 * np.pi, 3 * np.pi, 100)
```

```
y = np.sin(x)
plt.plot(x, y);
```

[22]: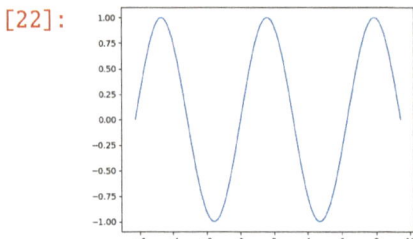

Here the sin function in the line y = np.sin(x) applies to each entry of the array x. This is known as a vector calculation. We will see this method in more detail later in this chapter and how it fits brilliantly with matplotlib.

Here are some more functions within numpy to generate arrays.

[23]: `np.zeros(3)`

[23]: `array([0., 0., 0.])`

[24]: `np.zeros((3, 6))`

[24]: `array([[0., 0., 0., 0., 0., 0.],`
`       [0., 0., 0., 0., 0., 0.],`
`       [0., 0., 0., 0., 0., 0.]])`

[25]: `z.shape`

[25]: `(3, 6)`

[26]: `np.zeros((3, 6, 4))`

[26]: `array([[[0., 0., 0., 0.],`
`        [0., 0., 0., 0.],`
`        [0., 0., 0., 0.],`
`        [0., 0., 0., 0.],`
`        [0., 0., 0., 0.],`
`        [0., 0., 0., 0.]],`

`       [[0., 0., 0., 0.],`
`        [0., 0., 0., 0.],`
`        [0., 0., 0., 0.],`
`        [0., 0., 0., 0.],`
`        [0., 0., 0., 0.],`

```
              [0., 0., 0., 0.]],

             [[0., 0., 0., 0.],
              [0., 0., 0., 0.],
              [0., 0., 0., 0.],
              [0., 0., 0., 0.],
              [0., 0., 0., 0.],
              [0., 0., 0., 0.]]])
```

`np.empty` creates a multi-dimensional array filled with random numbers. Here is a 5×5 array generated by this function.

```
[27]: np.empty([2**2 + 1,2**2 + 1])
```

```
[27]: array([[0.00000000e+000, 0.00000000e+000, 0.00000000e+000,
              0.00000000e+000, 2.12199579e-314],
             [1.15998412e-028, 4.31603868e-080, 1.94919985e-153,
              1.35717430e+131, 7.06652016e-096],
             [7.06652016e-096, 7.18988929e+140, 6.01347002e-154,
              6.98345625e-077, 6.98345624e-077],
             [6.98345624e-077, 6.01391519e-154, 7.06673073e-096,
              7.06652016e-096, 7.06652016e-096],
             [5.79961843e+294, 3.45365695e+175, 2.77191367e+296,
              2.91237123e+257, 4.71294503e+257]])
```

7.1.3 Accessing entries of an array

Accessing entries of an `array` is very similar to how we accessed elements of `lists` or `tuples`. The examples below show how this is done.

```
[28]: x = np.arange(1, 11)
```

```
[29]: x
```

```
[29]: array([ 1,  2,  3,  4,  5,  6,  7,  8,  9, 10])
```

```
[30]: x[0]
```

```
[30]: 1
```

```
[31]: x[-1]
```

```
[31]: 10
```

```
[32]: x[2 : 6]
```

```
[32]: array([3, 4, 5, 6])
```

```
[33]: x[4 : ]
```

```
[33]: array([ 5,  6,  7,  8,  9, 10])
```

```
[34]: x[ : 7]
```

```
[34]: array([1, 2, 3, 4, 5, 6, 7])
```

```
[35]: x[2 : 6] = 666
```

```
[36]: x
```

```
[36]: array([  1,   2, 666, 666, 666, 666,   7,   8,   9,  10])
```

```
[37]: mat = np.empty([2**2 + 1, 2**2 + 1])
```

```
[38]: mat[ : ] = 0
```

```
[39]: mat
```

```
[39]: array([[0., 0., 0., 0., 0.],
             [0., 0., 0., 0., 0.],
             [0., 0., 0., 0., 0.],
             [0., 0., 0., 0., 0.],
             [0., 0., 0., 0., 0.]])
```

```
[40]: mat[0, : ] = 666
```

```
[41]: mat
```

```
[41]: array([[666., 666., 666., 666., 666.],
             [  0.,   0.,   0.,   0.,   0.],
             [  0.,   0.,   0.,   0.,   0.],
             [  0.,   0.,   0.,   0.,   0.],
             [  0.,   0.,   0.,   0.,   0.]])
```

```
[42]: mat[-1, : ] = 666
```

```
[43]: mat
```

```
[43]:  array([[666., 666., 666., 666., 666.],
              [  0.,   0.,   0.,   0.,   0.],
              [  0.,   0.,   0.,   0.,   0.],
              [  0.,   0.,   0.,   0.,   0.],
              [666., 666., 666., 666., 666.]])
```

```
[44]:  mat[ : ,0] = 666
```

```
[45]:  mat
```

```
[45]:  array([[666., 666., 666., 666., 666.],
              [666.,   0.,   0.,   0.,   0.],
              [666.,   0.,   0.,   0.,   0.],
              [666.,   0.,   0.,   0.,   0.],
              [666., 666., 666., 666., 666.]])
```

```
[46]:  mat[ : , -1] = 666
```

```
[47]:  mat
```

```
[47]:  array([[666., 666., 666., 666., 666.],
              [666.,   0.,   0.,   0., 666.],
              [666.,   0.,   0.,   0., 666.],
              [666.,   0.,   0.,   0., 666.],
              [666., 666., 666., 666., 666.]])
```

We can do all the assignments in one go:

```
[48]:  mat[0, :] = mat[ :, 0] = mat[-1, : ] = mat[ :, -1] = 999
```

```
[49]:  mat
```

```
[49]:  array([[999., 999., 999., 999., 999.],
              [999.,   0.,   0.,   0., 999.],
              [999.,   0.,   0.,   0., 999.],
              [999.,   0.,   0.,   0., 999.],
              [999., 999., 999., 999., 999.]])
```

```
[50]:  mat[1 : 4, 1 : 4] = 555.0
```

```
[51]:  mat
```

```
[51]:  array([[999., 999., 999., 999., 999.],
              [999., 555., 555., 555., 999.],
              [999., 555., 555., 555., 999.],
```

```
        [999., 555., 555., 555., 999.],
        [999., 999., 999., 999., 999.]])
```

[52]: `xf = mat.astype(np.int64)`

[53]: `xf`

[53]:
```
array([[999, 999, 999, 999, 999],
       [999, 555, 555, 555, 999],
       [999, 555, 555, 555, 999],
       [999, 555, 555, 555, 999],
       [999, 999, 999, 999, 999]])
```

Exercise 7.1 *Generate the following matrix in* **numpy**

$$A = \begin{pmatrix} 1\ 2\ 3 \\ 4\ 5\ 6 \\ 7\ 8\ 9 \end{pmatrix}$$

and modify the entries to

$$A = \begin{pmatrix} 1\ 999\ 999 \\ 4\ 999\ 999 \\ 7\ \ 8\ \ \ 9 \end{pmatrix}$$

Solution

First things first, we need to define the matrix using **numpy**'s `array`. This is a small matrix and we could just enter the entries by hand.

[54]: `x = np.array([[1, 2, 3], [4, 5, 6], [7, 8, 9]])`

[55]: `x`

[55]:
```
array([[1, 2, 3],
       [4, 5, 6],
       [7, 8, 9]])
```

Here is yet another way to define this matrix via list comprehension.

[56]: `t = [[j for j in range(1 + i, 4 + i)] for i in range(0, 9,`
      `↪3)]`

[57]: `x = np.array(t)`

[58]: `x`

```
[58]:  array([[1, 2, 3],
               [4, 5, 6],
               [7, 8, 9]])
```

```
[59]:  print(x[0][0], x[1][0], x[2][0])
```

```
       1 4 7
```

```
[60]:  x[ : ,0]
```

```
[60]:  array([1, 4, 7])
```

```
[61]:  x[ : ,1]
```

```
[61]:  array([2, 5, 8])
```

```
[62]:  x[ :2, 1:]
```

```
[62]:  array([[2, 3],
               [5, 6]])
```

Here we ask Python to grab all the entries up to the second item from the array by :2, so, [1, 2, 3], and [4, 5, 6]. Next by 1: we grab all the entries starting from 1, so [2, 3], and [5, 6].

```
[63]:  x[ :2, 1:] = 999
```

```
[64]:  x
```

```
[64]:  array([[  1, 999, 999],
               [  4, 999, 999],
               [  7,   8,   9]])
```

There is one difference between assigning values to np.array and lists. The following example will show this:

```
[65]:  x = np.arange(5, 17)
```

```
[66]:  x
```

```
[66]:  array([ 5,  6,  7,  8,  9, 10, 11, 12, 13, 14, 15, 16])
```

```
[67]:  y = x[3 : 8]
```

```
[68]:  y
```

```
[68]: array([ 8,   9, 10, 11, 12])
```

```
[69]: y[ : ] = 999
```

```
[70]: y
```

```
[70]: array([999, 999, 999, 999, 999])
```

```
[71]: x
```

```
[71]: array([  5,    6,    7, 999, 999, 999, 999, 999,  13,   14,   15,
       16])
```

This shows that when we specify y = x[3 : 8], then y is still pointing to the portion of the same object as x. Thus changing y would change x as well.

To create a new array, one can use the method .copy().

```
[72]: z = x[3 : 8].copy()
```

```
[73]: z[ : ] = -1
```

```
[74]: z
```

```
[74]: array([-1, -1, -1, -1, -1])
```

```
[75]: x
```

```
[75]: array([  5,    6,    7, 999, 999, 999, 999, 999,  13,   14,   15,
       16])
```

This is the difference with lists. The example below shows the behaviour of lists.

```
[76]: t = [5, 6, 7, 8, 9, 10, 11, 12, 13, 14, 15, 16]
```

```
[77]: t
```

```
[77]: [5, 6, 7, 8, 9, 10, 11, 12, 13, 14, 15, 16]
```

```
[78]: z = t[3 : 8]
```

```
[79]: z
```

```
[79]: [8, 9, 10, 11, 12]
```

```
[80]: z[1] = 777
```

```
[81]: z
```

```
[81]: [8, 777, 10, 11, 12]
```

```
[82]: t
```

```
[82]: [5, 6, 7, 8, 9, 10, 11, 12, 13, 14, 15, 16]
```

7.1.4 Vector calculus with arrays

numpy has "all" the tools available for doing linear algebra, such as determinant, eigenvalues, eigenvectors and many more.

One of the questions asked earlier when we were working with lists was the following: Given $\mathbf{x} = \{x_1, x_2, \cdots, x_n\}$ and $\mathbf{y} = \{y_1, y_2, \cdots, y_n\}$, how can one produce

$$\{x_1 + y_1, x_2 + y_2, \cdots, x_n + y_n\}?$$

One approach is to consider lists as np.array x = array([x1, x2, ..., xn)] and y = array([y1, y2, ..., yn)], then the sum of vectors x + y is what we want. Then, as you might also guess, x * y would produce array([x1 y1, x2 y2, ...,xn yn]), i.e., all arithmetical operations performed on array are component-wise.

However, there is another product in the setting of vectors, namely the *inner product*, which is defined as

$$\mathbf{x}.\mathbf{y} = x_1y_1 + x_2y_2 + \cdots + x_ny_n.$$

The following shows how numpy handles these different operations.

```
[83]: x = np.arange(2, 7); y = np.arange(7, 2, -1)
```

```
[84]: print('', x,'\n', y)
```

```
[84]:  [2 3 4 5 6]
       [7 6 5 4 3]
```

```
[85]: x + y
```

```
[85]: array([9, 9, 9, 9, 9])
```

```
[86]: x * y
```

```
[86]: array([14, 18, 20, 20, 18])
```

```
[87]: x ** y == [2**7, 3**6, 4**5, 5**6, 6**3]
```

```
[87]: array([ True,   True,   True, False,   True])
```

It is known that matrix calculation is a tedious job. It will take well over 10 minutes to multiply

$$
\begin{pmatrix}
2 & -3 & 13 & -4 & 8 \\
12 & 1 & -18 & -4 & 2 \\
18 & 21 & 10 & 0 & 9 \\
8 & -12 & -4 & 0 & -3 \\
15 & -7 & 2 & 4 & 2
\end{pmatrix}
\times
\begin{pmatrix}
11 & 34 & -21 & 0 & -43 \\
12 & -33 & 9 & -12 & 7 \\
16 & -7 & -43 & 84 & 3 \\
4 & 9 & 12 & -1 & -54 \\
7 & 22 & -5 & 23 & 0
\end{pmatrix}
$$

by hand, most likely only to obtain a wrong answer!

We generate a 3×2 matrix A defined by (a_{ij}) with entries $a_{ij} = i - j$ and a 2×3 matrix B defined by $b_{ij} = i + j^2$ and calculate the matrix products $A.B$ and $B.A$ in numpy; among other things this will show that matrix multiplication is not commutative.

We define the matrix A first, using list comprehension.

```
[88]: t = [[i - j for j in range(1, 3)] for i in range(1, 4)]

      A = np.array(t)
```

```
[89]: A
```

```
[89]: array([[ 0, -1],
             [ 1,  0],
             [ 2,  1]])
```

```
[90]: s = [[i + j**2 for j in range(1, 4)] for i in range(1, 3)]

      B = np.array(s)
```

```
[91]: B
```

```
[91]: array([[ 2,  5, 10],
             [ 3,  6, 11]])
```

For matrix multiplication, we use the method dot.

```
[92]:  np.dot(A,B)
```

```
[92]:  array([[ -3,  -6, -11],
              [  2,   5,  10],
              [  7,  16,  31]])
```

Or using the method in the object:

```
[93]:  A.dot(B)
```

```
[93]:  array([[ -3,  -6, -11],
              [  2,   5,  10],
              [  7,  16,  31]])
```

```
[94]:  B.dot(A)
```

```
[94]:  array([[25,  8],
              [28,  8]])
```

Exercise 7.2 *Consider the system of equations*

$$\begin{cases} 3x + 2y = 3 \\ 10x - y = 5. \end{cases}$$

Find the unique solution to these equations.

Solution

Representing this system via the language of matrices points to a natural way to use matrix calculus to solve the equations. Writing

$$\begin{pmatrix} 3 & 2 \\ 10 & -1 \end{pmatrix} \begin{pmatrix} x \\ y \end{pmatrix} = \begin{pmatrix} 3 \\ 5 \end{pmatrix}$$

then clearly

$$\begin{pmatrix} x \\ y \end{pmatrix} = \begin{pmatrix} 3 & 2 \\ 10 & -1 \end{pmatrix}^{-1} \begin{pmatrix} 3 \\ 5 \end{pmatrix}.$$

Now we use numpy to compute the inverse of the matrix and get the result.

```
[95]:  import numpy as np

       A = np.array([[3, 2], [10, -1]])
       B = np.array([3, 5])
```

The library numpy provides a method to calculate the inverse of non-singular matrices.

```
[96]: inv_A = np.linalg.inv(A)

      print(inv_A)
```

```
[96]: [[ 0.04347826  0.08695652]
       [ 0.43478261 -0.13043478]]
```

So the solution to this system of equations is:

```
[97]: print(np.dot(inv_A, B))
```

```
[97]: [0.56521739 0.65217391]
```

Or we could use the matrix object's dot method and do all these in one line:

```
[98]:  np.linalg.inv(A).dot(B)
```

```
[98]: array([0.56521739, 0.65217391])
```

In fact numpy also provides a method to solve the system of equations directly.

```
[99]: np.linalg.solve(A,B)
```

```
[99]: array([0.56521739, 0.65217391])
```

Getting help from sympy graphics from Chapter 6, we plot the lines $3x + y = 4$ and $10x - y = 5$ and observe that they indeed intersect at one point.

```
[100]: from sympy import symbols, Eq
       from sympy import plot_implicit

       x, y = symbols('x y')

       p1 = plot_implicit(Eq(3 * x + 2 * y, 3),  (x, -2, 2), (y, -2,
       ↪2),line_color='blue', show=False);
       p2 = plot_implicit(Eq(10 * x - y, 5),  (x, -2, 2), (y, -2,
       ↪2),line_color='green', show=False);
       p1.append(p2[0])
       p1.show()
```

[100]:
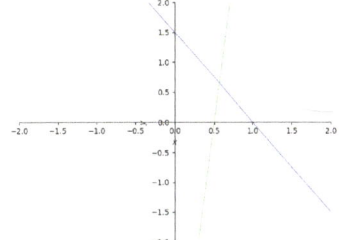

A remark is in order. We did multiply a 2×2 matrix A with 2×1 matrix B. So we should have defined B as B = np.array([[3], [5]]). However we simply defined B = np.array([3, 5]) and numpy did consider this as a column and the multiplications was valid.

Exercise 7.3 *Consider the system of equations*

$$\begin{cases} 3x + 2y - 4z = 2 \\ 3x + 2y - (4 + R)z = \frac{1}{2} \\ 3x + (2 + S)y - 4z = 3. \end{cases}$$

First find the unique solution for this system when $R = 0.05$ and $S = -0.05$. Next explore if R and S change, how the solution behaves, and how far the new solution would be from initial solution.

Solution

This interesting problem was investigated by Shiskowski and Frinkle (Example 3.3.5) in their book *Principles of Linear Algebra with Mathematica*, Wiley & Sons, 2011.

We first define the coefficient matrix and the value matrix in order to use numpy's linear system solve method.

```
101]: import numpy as np

def sys(r,s):
    A = np.array([[3, 2, -4], [3, 2, -4 - r], [3, 2 + s,
    ↪-4]])
    return A

p = np.array([2, 1/2, 3])

sys(0.05, -0.05)
```

```
101]: array([[ 3.,  2.  ,  -4.  ],
             [ 3.,  2.  ,  -4.05],
             [ 3.,  1.95, -4.  ]])
```

```
[102]: init_sol = np.linalg.solve(sys(0.05, -0.05), p)

       init_sol
```

[102]: `array([ 54., -20.,  30.])`

If we change the s slightly, the solutions change substantially.

```
[103]: np.linalg.solve(sys(0.01, -0.1), p)
```

[103]: `array([207.33333333, -10., 150. ])`

The following graph explains why this system of equations is so sensitive to changes of R and S. The planes are almost parallel, so a slight change to R or S will move the intersection of the planes by a large margin.

```
[104]: from sympy.plotting import plot3d
       from sympy import symbols

       def p1(x, y):
           return (3 * x + 2 * y - 2) / 4

       def p2(x, y, r):
           return (3 * x + 2 * y - 1/2) / (4 + r)

       def p3(x, y, s):
           return (3 * x + (2 + s) * y - 3) / 4

       x, y = symbols('x y')

       plot3d(p1(x, y), p2(x, y, 0.05), p3(x, y, -0.05), (x, -2, 2),␣
        ↪(y, -2, 2));
```

[104]:

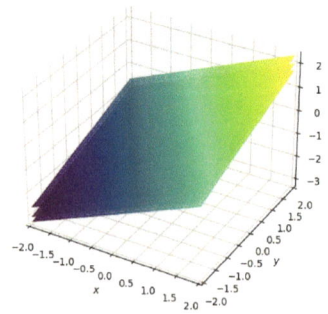

The distance between two points (x_1, y_1, z_1) and (x_2, y_2, z_2) in space is measured by

$$\sqrt{(x_1 - y_1)^2 + (x_2 - y_2)^2 + (x_3 - y_3)^2}.$$

The numpy norm method can be used to get this formula.

```
[105]: import numpy as np

       np.linalg.norm(np.array([2,0,0]))
```

```
[105]: 2.0
```

Putting all these codes together:

```
[106]: def dis_fuc(r,s):
           init_sol = np.array([ 54., -20.,   30.])
           rs_sol = np.linalg.solve(sys(r, s), p)
           dis = np.linalg.norm(init_sol - rs_sol)
           return dis

       dis_fuc(1,2)
```

```
[106]: 62.46554605896313
```

Exercise 7.4 *Define the $n \times n$ matrix $A_n = (a_{ij})$ whose i, j-th entries are*

$$a_{ij} = \begin{cases} 1 & i = j \\ i^2 + j^2 & i \neq j. \end{cases}$$

Show that for $1 \leq n \leq 10$ the determinant of A_n is negative for odd values and positive for even values of n.

Solution

We first define a function that, for each n, will generate the matrix A_n. Then we use the determinant function within numpy to evaluate the determinant of A_n. The function numpy.linalg.det evaluates the determinant of a matrix.

```
[107]: def A(n):
           mat = np.empty([n, n])
           for i in range(n):
               for j in range(n):
                   if i == j:
                       mat[i, i] = 1
                   else:
                       mat[i, j] = (i + 1)**2 + (j + 1)**2
           return mat
```

```
from numpy.linalg import det

det_odd = {'even indices': [round(det(A(n))) for n in
    ↪range(1, 10, 2)]}
det_even = {'odd indices': [round(det(A(n))) for n in
    ↪range(2, 10, 2)]}

print(det_odd,'\n' '\n', det_even)
```

[107]: {'even indices': [1, 1007, 5771871, 102458829221,
4336469213954120]}

{'odd indices': [-24, -64591, -681222836, -19122185858615]}

Note that the function A(n) contains a local variable mat. The initial value of this variable is the $n \times n$ array defined by mat = np.empty([n, n]). We then fill the entries of this array as described by the matrix A_n.

In the next exercise, we define another matrix. This time we first use a list comprehension to define the matrix and then convert it into array within the numpy library.

Exercise 7.5 *Let B_n denote the $n \times n$ matrix with (i, j)-th entry equal to*

$$
b_{ij} = \begin{cases}
\frac{1}{2j-i^2} & \text{if } i > j \\[2mm]
\frac{1}{i-j} + \frac{1}{n^2-j-i} & \text{if } j > i \\[2mm]
0 & \text{if } i = j.
\end{cases}
$$

Define a function $B(n)$ to generate this matrix for any n. Look at the numerical values of the determinant of B_n for $3 \leq n \leq 15$.

Solution

We define the function B(n) which generates the above matrix for any n. We first use list comprehension to define the matrix. We give an alternative code using a nested for-loop as well.

[108]:
```
def B(n):
    t = [[1/(2*j - i**2) if i>j
          else 1/(i - j) + 1/(n**2 - j - i) if j>i
          else 0
```

```
        for j in range(1, n+1)] for i in range(1, n+1)]
    return t
```

[109]: `np.array(B(5))`

[109]:
```
array([[ 0.,  -0.95454545, -0.45238095, -0.28333333,
          -0.19736842],
       [-0.5,   0.,  -0.95, -0.44736842, -0.27777778],
       [-0.14285714, -0.2,   0.,  -0.94444444, -0.44117647],
       [-0.07142857, -0.08333333, -0.1,   0.,  -0.9375],
       [-0.04347826, -0.04761905, -0.05263158, -0.05882353,
          0. ]])
```

The function numpy.linalg.det evaluates the determinant of a matrix.

[110]: `from numpy.linalg import det`

[111]: `[det(B(n)) for n in range(3,11)]`

[111]:
```
[-0.11928571428571424,
 -0.03525757631818237,
 -0.018062205431347532,
 -0.013023479578175417,
 -0.009958843179948978,
 -0.007821845590148313,
 -0.0062884030416988125,
 -0.005158398365984703]
```

Next we generate the matrix B(n) using a nested for-loop. Here the first loop for i will generate the rows and for j generates the columns. We then generate each entry aij of each row and we complete each row via row = row + aij. Once each row is ready then, via b = b + [row], we collect the rows into the matrix b.

[112]:
```
def B(n):
    b = []
    for i in range(1, n+1): # generating i-th row
        row = []
        for j in range(1, n+1):  # generating j-th column
            if (i > j):
                row += [1 / (2*j - i**2)]  # adding bij entry␣
   ↪to the row
            elif (i < j):
                row += [(1 / (i-j)) + (1 / (n**2 - j - i))]
            else:
                row += [0]
```

```
            b += [row]
        return b
```

[113]: `np.array(B(5))`

[113]:
```
array([[ 0.,       -0.95454545, -0.45238095, -0.28333333,
          -0.19736842],
        [-0.5,      0.,        -0.95,       -0.44736842, -0.27777778],
        [-0.14285714, -0.2,      0.,        -0.94444444, -0.44117647],
        [-0.07142857, -0.08333333, -0.1,      0.,        -0.9375],
        [-0.04347826, -0.04761905, -0.05263158, -0.05882353,
          0. ]])
```

Once we have the determinant, it is easy to observe its behaviour on a graph. We will return to a similar example as a case study in Chapter 8.

[114]: `import matplotlib.pyplot as plt`

[115]:
```
det_list = [det(B(i)) for i in range(1, 15)]

plt.plot(det_list, 'b--');
```

[115]: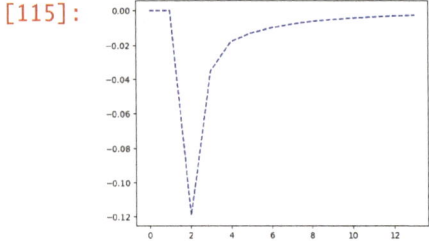

Exercise 7.6 *Define a matrix as follows:*

$$d(n) = \begin{pmatrix} 1 & 2 & \cdots & n \\ n+1 & n+2 & \cdots & 2n \\ \vdots & \vdots & \vdots & \vdots \\ \cdots & \cdots & \cdots & n^2 \end{pmatrix}.$$

Check that, for any $2 < n < 15$, $\det(d(n)) = 0$.

Solution

We try several different ways to generate this matrix. We first start with list comprehension, then nested for-loops and then we give an approach using numpy's method reshape.

```
[116]: def A(n):
           s = [[i for i in range(1 + n*j, n+1 + n*j)] for j in
           ↪range(0, n)]
           return s
```

```
[117]: A(3)
```

```
[117]: [[1, 2, 3], [4, 5, 6], [7, 8, 9]]
```

```
[118]: A(7)
```

```
[118]: [[1, 2, 3, 4, 5, 6, 7],
        [8, 9, 10, 11, 12, 13, 14],
        [15, 16, 17, 18, 19, 20, 21],
        [22, 23, 24, 25, 26, 27, 28],
        [29, 30, 31, 32, 33, 34, 35],
        [36, 37, 38, 39, 40, 41, 42],
        [43, 44, 45, 46, 47, 48, 49]]
```

We give an alternative to generate this matrix via a for-loop.

```
[119]: def A1(n):
           A = []
           for j in range(0, n):
               row = []
               for i in range(1 + n * j, n + 1 + n * j):
                   row += [i]
               A += [row]
           return A
```

```
[120]: A1(7)
```

```
[120]: [[1, 2, 3, 4, 5, 6, 7],
        [8, 9, 10, 11, 12, 13, 14],
        [15, 16, 17, 18, 19, 20, 21],
        [22, 23, 24, 25, 26, 27, 28],
        [29, 30, 31, 32, 33, 34, 35],
        [36, 37, 38, 39, 40, 41, 42],
        [43, 44, 45, 46, 47, 48, 49]]
```

And finally, the method reshape provides the easiest way to generate this matrix.

```
[121]: A = np.arange(1, 50).reshape(7, 7)
```

```
[122]: A
```

```
[122]: array([[ 1,  2,  3,  4,  5,  6,  7],
              [ 8,  9, 10, 11, 12, 13, 14],
              [15, 16, 17, 18, 19, 20, 21],
              [22, 23, 24, 25, 26, 27, 28],
              [29, 30, 31, 32, 33, 34, 35],
              [36, 37, 38, 39, 40, 41, 42],
              [43, 44, 45, 46, 47, 48, 49]])
```

```
[123]: def A2(n):
           return np.arange(1, n**2 + 1).reshape(n, n)
```

```
[124]: A2(3)
```

```
[124]: array([[1, 2, 3],
              [4, 5, 6],
              [7, 8, 9]])
```

Now we are ready to calculate the determinant of the matrix.

```
[125]: from numpy.linalg import det

       [round(det(A2(n))) for n in range(3,16)]
```

```
[125]: [0, 0, 0, 0, 0, 0, 0, 0, 0, 0, 0, 0, 0]
```

Exercise 7.7 *Write a function to accept a matrix A_{nn} and produce the $n^2 \times n^2$ matrix B as follows,*

$$
\left(
\begin{array}{cccc}
\begin{pmatrix} a_{11} & 0 & 0 \\ 0 & \ddots & 0 \\ 0 & 0 & a_{11} \end{pmatrix} & \begin{pmatrix} a_{12} & 0 & 0 \\ 0 & \ddots & 0 \\ 0 & 0 & a_{12} \end{pmatrix} & \cdots & \begin{pmatrix} a_{1n} & 0 & 0 \\ 0 & \ddots & 0 \\ 0 & 0 & a_{1n} \end{pmatrix} \\
\vdots & \vdots & \vdots & \vdots \\
\begin{pmatrix} a_{n1} & 0 & 0 \\ 0 & \ddots & 0 \\ 0 & 0 & a_{n1} \end{pmatrix} & \cdots & \cdots & \begin{pmatrix} a_{nn} & 0 & 0 \\ 0 & \ddots & 0 \\ 0 & 0 & a_{nn} \end{pmatrix}
\end{array}
\right).
$$

Then show that $\det(A)^n = \det(B)$.

Solution

We write the program for a 3×3 matrix. It is very easy to adapt this to an $n \times n$-matrix. The challenge is to get the indices of the entries right. We design three nested-loops, *k* for moving down the columns, *i* for moving right in each row and *j* to fill the diagonals with the given entries of the matrix *A*.

```
[268]: A = np.array([[1, 2, 3],
                     [4, 5, 6],
                     [7, 8, 9]])

       B = np.zeros((9, 9))

       for k in range(3):
           for i in range(3):
               for j in range(3):
                   B[j + k * 3, i * 3 + j] = A[k, i]
```

```
[127]: B
```

```
[127]: array([[1., 0., 0., 2., 0., 0., 3., 0., 0.],
              [0., 1., 0., 0., 2., 0., 0., 3., 0.],
              [0., 0., 1., 0., 0., 2., 0., 0., 3.],
              [4., 0., 0., 5., 0., 0., 6., 0., 0.],
              [0., 4., 0., 0., 5., 0., 0., 6., 0.],
              [0., 0., 4., 0., 0., 5., 0., 0., 6.],
              [7., 0., 0., 8., 0., 0., 9., 0., 0.],
              [0., 7., 0., 0., 8., 0., 0., 9., 0.],
              [0., 0., 7., 0., 0., 8., 0., 0., 9.]])
```

```
[128]: from numpy.linalg import det

       round(det(A)**3, 7) == round(det(B),7)
```

```
[128]: True
```

We finish this section by looking at a very useful function available in numpy, namely meshgrid. One of the applications of this function is to create a contour plot of a graph. This we will do in Chapter 8. Here we explain how meshgrid works.

```
[129]: import numpy as np

       x = [1, 2, 3]
       y = [10, 11]
       np.meshgrid(x, y)
```

```
[129]: [array([[1, 2, 3],
               [1, 2, 3]]),
```

```
array([[10, 10, 10],
       [11, 11, 11]])]
```

Suppose $x = (x_1, x_2, \cdots, x_m)$ and $y = (y_1, y_2, \cdots, y_n)$. Then np.meshgrid(x, y) gives two matrices of the form

```
[[array([[x_1, x_2, ..., x_m],
         [x_1, x_2, ..., x_m],
         [x_1, x_2, ..., x_m],
         ................. ,
         [x_1, x_2, ..., x_m]),
  array([[y_1, y_1, ..., ., y_1],
         [y_2, y_2, ..., ., y_2],
         [.  ,  . , ..., ., . ],
         [.  ,  . , ..., ., . ],
         [y_n, y_n, ., ., y_n]])]
```

A close inspection shows the first matrix comprises n rows, each a copy of x, and the second matrix comprises m columns, each a copy of y. To understand this better, one can look at the following diagram. The meshgrid for $x = (x_1, x_2)$ and $y = (y_1, y_2)$ is depicted as a three-dimensional function on the X-Y-plane.

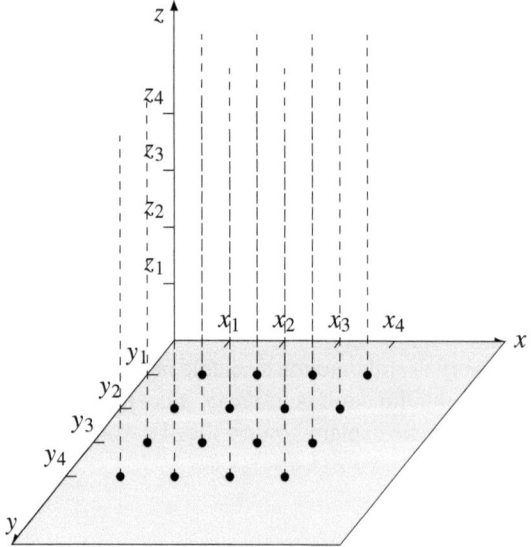

7.2 Universal Functions

When working with `arrays`, the built-in `numpy` functions behave similarly to the way functions are mapped to lists, that is, the functions are applied to each entry of the `array`. This allows for powerful and yet short codes.

As an example, consider the following two `np.array`

$$`x = array([x1, x2, ..., xn)]`$$
$$`y = array([y1, y2, ..., yn)]`$$

Making the functions `sin` and `cos` available by importing them from the `numpy` library, the expression `sin(x) + cos(y**2)` produces

```
array([sin(x1)+cos(y1**2), sin(x2)+cos(y2**2), ...,
       sin(xn)+cos(yn**2)]).
```

Writing mathematically, given $x = (x_1, x_2, \ldots, x_n)$ and $y = (y_1, y_2, \ldots, y_n)$, the expression $\sin(x) + \cos(y^2)$ generates

$$\left(\sin(x_1) + \cos(y_1^2), \ \sin(x_2) + \cos(y_2^2), \ \cdots, \ \sin(x_n) + \cos(y_n^2) \right).$$

So the reader can recognise the conceptual shift to "vectorisation" computation in `numpy`, as opposed to arithmetic with `lists`.

```
[130]: x = np.arange(1, 11)
```

We compute $2^i - 1$, for $1 \leq i \leq 10$.

```
[131]: print('', x, '\n\n', 2**x - 1)
```

```
[131]:  [ 1  2  3  4  5  6  7  8  9 10 ]
        [ 1  3  7 15 31 63 127 255 511 1023 ]
```

Exercise 7.8 *Using numpy demonstrate that*

$$\frac{1 + \sin(x) - \cos(x)}{1 + \sin(x) + \cos(x)} = \tan(x/2).$$

Solution

We define the right- and left-hand side of this identity as two functions, evaluate them for variety of data, and then compare the results.

```
[132]: def npright_side(x):
           return (1 + np.sin(x) - np.cos(x))/(1 + np.sin(x) + np.
       ↪cos(x))
```

```
def npleft_side(x):
    return np.tan(x/2)
```

[133]: x = np.linspace(0, 2 * np.pi, 20)
        npright_side(x) - npright_side(x)

[133]: array([0., 0., 0., 0., 0., 0., 0., 0., 0., 0., 0., 0., 0., 0.,
               0., 0., 0., 0., 0., 0.])

Problems

1. Given two arrays $\{x_1, x_2, \cdots, x_n\}$ and $\{y_1, y_2, \cdots, y_n\}$, in numpy, produce the following arrays:

   - $\{x_1, y_1, x_2, y_2, \cdots, x_n, y_n\}$,
   - $\{\{x_1, y_1\}, \{x_2, y_2\}, \cdots, \{x_n, y_n\}\}$,
   - $\{x_1 + y_1, x_2 + y_2, \cdots, x_n + y_n\}$,
   - $\{x_1, x_1 + x_2, \cdots, x_1 + x_2 + \cdots + x_n\}$,
   - $\left\{\{\{x_1\}, \{x_2, \ldots, x_n\}\}, \{\{x_1, x_2\}, \{x_3, \ldots, x_n\}\} \ldots \{\{x_1, \ldots x_{n-1}\}, \{x_n\}\}\right\}$,
   - and,

   $$\{\{x_1, y_1\}, \{x_1, y_2\}, \cdots, \{x_1, y_n\}, \{x_2, y_1\}, \{x_2, y_2\}, \cdots, \{x_2, y_n\}, \cdots,$$
   $$\{x_n, y_1\}, \{x_n, y_2\}, \cdots, \{x_n, y_n\}\}.$$

2. The matrix $e_{i,j}^n(a)$ is defined as an $n \times n$ matrix with ones on the diagonal, a in the i-th row and j-th column and 0 everywhere else, for example:

   $$e_{1,2}^3(a) = \begin{pmatrix} 1 & a & 0 \\ 0 & 1 & 0 \\ 0 & 0 & 1 \end{pmatrix}$$

   Write a function to create $e_{i,j}^n(a)$. Then show that

   - $\det(e_{3,4}^7(a)) = 1$

   - $e_{3,4}^7(a).e_{3,4}^7(b) = e_{3,4}^7(a + b)$

- Let A be an arbitrary 5×5 matrix. Show that $e^5_{3,4}(a).A$ adds the multiple a of the 4-th row of A to the 3-rd row of A.

  Furthermore define a function g which accepts two $n \times n$ matrices A and B by $g(A, B) = A.B.A^{-1}B^{-1}$.

  Then show that

- $g(e^7_{3,4}(a), e^7_{4,6}(b)) = e^7_{3,6}(ab)$.

3. Let B_n denote the $n \times n$-matrix with (i, j)-th entry equal to $1/(|i-j|)$ if $i \neq j$ and 0 if $i = j$. Find the value of the determinant of B_n for all values of $1 \leq n \leq 20$.

4. Define the $n \times n$ matrix $A_n = (a_{ij})$ whose i, j-th entries are

$$a_{ij} = \begin{cases} i + j & \text{if } i = j \\ i^{i+j} & \text{if } i \neq j. \end{cases}$$

   Find out how long it takes Python to evaluate the inverse of $A(n)$ for $n = 15, 16, \ldots 30$ (this might take some seconds).

5. Consider a 10×10 matrix with positive integers as its entries. Write a function θ which is the sum of the third largest number of each row.

   Now create a 10×10 matrix M whose entries are the integers 1 to 100:

$$M = \begin{pmatrix} 1 & 2 & \cdots & 10 \\ 11 & 12 & \cdots & 20 \\ \vdots & \vdots & \vdots & \vdots \\ 91 & 92 & \cdots & 100 \end{pmatrix}.$$

   Show that $\theta(M)$ is greater than the sum of the numbers in some row.

Chapter 8
The `matplotlib` Library and Projects

8.1 `matplotlib`, Plotting Data

In this chapter we will look more closely at the Python graphics library `matplotlib`, which is widely used for plotting data and creating professional and magnificent two-dimensional graphics. This library works perfectly with `numpy`. With `numpy` we handle the data and with `matplotlib` we visualise them.

```
[1]: import matplotlib.pyplot as plt
```

Let us remind ourselves of the most basic way we can use the library.

```
[2]: plt.plot([1, 2, 2.5, 4], [10, 12, -2, 1]);
```

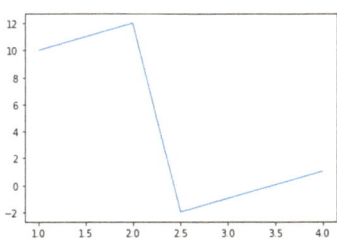

As the example shows, in order to plot a graph, we specify the x-coordinates and then the y-coordinates. If $x = (x_1, x_2, \cdots, x_n)$ and $y = (y_1, y_2, \cdots, y_n)$, then `plt.plot(x,y)` will produce a graph determined by the pairs $(x_i, y_i), 1 \leq i \leq n$.

We can introduce styles and formatting to our graphs. Here is just one example. The option `b.` displays small blue filled circles and `r^` displays solid red triangles pointing upward.

© The Author(s), under exclusive license to Springer Nature Switzerland AG 2023
R. Hazrat, *A Course in Python*, Springer Undergraduate Mathematics Series,
https://doi.org/10.1007/978-3-031-49780-3_8

```
[3]: plt.plot(range(1, 20), range(1,20), 'b.');
     plt.plot(range(20, 1, -1), range(1,20), 'r^');
     plt.plot(range(1, 21), [10 for i in range(1, 21)],↵
        ↪color='green', marker='o', linestyle='dashed');
```

[3]:

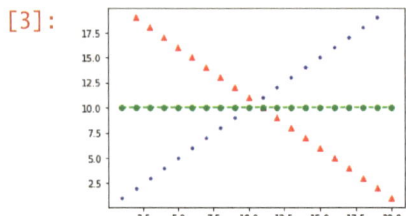

Staying with lists, we calculate the sin function for certain values and plot the result.

```
[4]: import math

     x = range(-10, 10)
     y = [math.sin(i) for i in x]
```

```
[5]: plt.plot(x, y, 'g^-');
```

[5]:

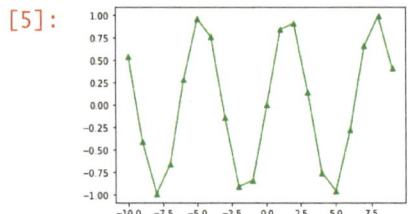

We won't spend too much time on how to style the outputs. The examples below show some samples and the codes for any combinations and decorations one has in mind are just a google away.

As mentioned, `matplotlib` works perfectly with `numpy`. This allows us to do all kind of calculations with ease, since `numpy` allows the use of universal functions, i.e. we can apply functions to each component of the array.

The example below produces the plot of the function $\sin(x)/x$, where $-10 \le x \le 10$.

```
[6]: import numpy as np
     import matplotlib.pyplot as plt

     x = np.arange(-10, 10, 0.2)
     plt.plot(x, np.sin(x)/x, 'g-^');
```

[6]: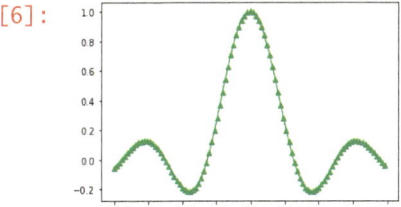

Exercise 8.1 *Plot the graph of* $f(x) = \frac{\sin(x)}{\cos(x)}$, *for* $0 \le x \le 10\pi$.

Solution

The universal function facility of numpy allows us to treat an array as single data. We first define an array $x = (x_1, x_2, \cdots, x_n)$, then when using numpy's sin and cos, the expression $y = \sin(x)/\cos(x)$ gives

$$y = \left(\frac{\sin(x_1)}{\cos(x_1)}, \frac{\sin(x_2)}{\cos(x_2)}, \cdots, \frac{\sin(x_n)}{\cos(x_n)} \right).$$

We can then simply pass y into the plot command.

```
[7]: x = np.arange(0.01, 10*np.pi, 0.1)
     y = np.sin(x)/np.cos(x)

     plt.plot(x, y);
```

[7]: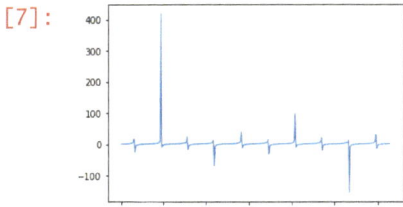

matplotlib allows us to plot several equations at the same time. For this one introduces the equations into plot by using a list containing all the equations.

As an example we plot the graphs of the functions $\sin(\frac{1}{x^2-x})$ and $\cos(\frac{1}{x^2-x})$ in the range $[0, \pi]$.

```
[8]: x = np.arange(np.pi/4, np.pi, 0.005)

     y = np.sin(1/(x**2 - x))
     z = np.cos(1/(x**2 - x))

     plt.plot(x, y, x, z);
```

[8]: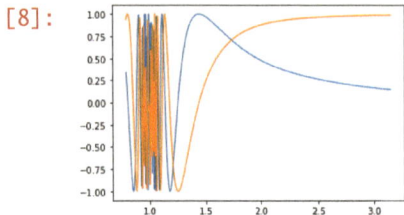

We could also use the function plot separately.

[9]:
```python
plt.plot(x, y);
plt.plot(x, z);
```

[9]: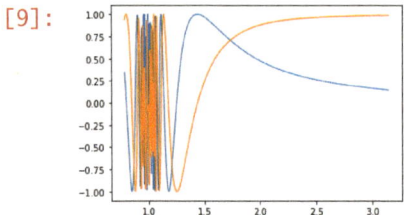

The way we can think of this is that plt.plot creates an object and each time we feed it the coordinates, we create new AA figure object can have several Axes, and Python then shows all of them together in one figure.

Exercise 8.2 *Plot the graphs of*

$$f_n(x) = \frac{\sin(x) - \cos(nx)}{1+x}$$

between 0 and 7π as n ranges from 1 to 10.

Solution

We define the function f_n in Python, which depends on n. We then create a loop, running n from 0 to 9.

[10]:
```python
import numpy as np
import matplotlib.pyplot as plt

x = np.arange(0, 7*np.pi, 0.005)

def f(n):
    return (np.sin(x) - np.cos(n * x))/(1+x)

for n in range(10):
    plt.plot(x, f(n));
```

[10]: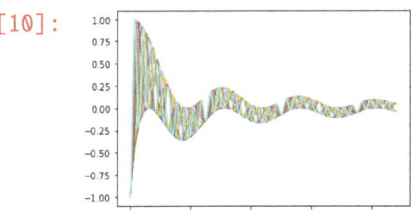

As one can see, although the `plt.plot` is repeated 10 times, all the graphs appear in one figure, as the plot defines a single object consisting of all these coordinates.

One can see that there is no difference between plotting a parametric plot and functions of the form $f(x) = \sin(x)$. Here, we plot the parametric function $x = \sin(t)$, $y = \cos(t)$.

```
[11]: t = np.linspace(-np.pi, np.pi, 400)
      x = np.sin(t)
      y = np.cos(t)
      plt.plot(x,y);
```

[11]: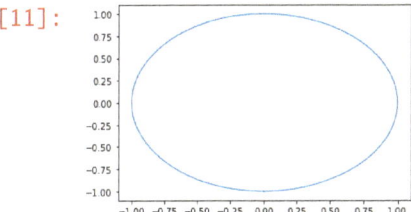

We expect to get a perfect circle. A closer look shows Python has used different scales for x and y-axis. We adjust this:

```
[12]: plt.axis('equal')
      plt.plot(x,y);
```

[12]:

```
[13]: plt.axis('equal')
      for i in range(10):
          plt.plot(x*i + i ,y*i);
```

[13]: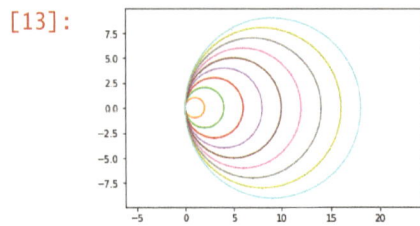

We have plotted the butterfly curve given by

$$x(t) = \sin(t)\left(e^{\cos(t)} - 2\cos(4t) - \sin^5(t/12)\right)$$

$$y(t) = \cos(t)\left(e^{\cos(t)} - 2\cos(4t) - \sin^5(t/12)\right)$$

using sympy graphics. Here we will plot this parametric equation using matplotlib.

[14]:
```
def x(t):
    return np.sin(t)*(np.exp(np.cos(t)) - 2*np.cos(4*t) - np.
    ↪sin(t/12)**5)
```

[15]:
```
def y(t):
    return np.cos(t)* (np.exp(np.cos(t)) - 2*np.cos(4*t) -
    ↪np.sin(t/12)**5)
```

[16]:
```
import numpy as np
import matplotlib.pyplot as plt

x_axis = x(np.linspace(-50, 50, 40000))
y_axis = y(np.linspace(-50, 50, 40000))
```

[17]:
```
plt.figure(figsize=(4, 4));
plt.axis('off');
plt.plot(x_axis, y_axis, linewidth=0.2);
```

[17]:

Exercise 8.3 *Plot the graph of*

$$x(t) = \cos(t) + 1/2\cos(7t) + 1/3\sin(17t)$$
$$y(t) = \sin(t) + 1/2\sin(7t) + 1/3\cos(17t)$$

for $0 \le t \le 14\pi$.

Solution

This is very similar to the previous exercise. We generate the x and y-coordinates and then use plot to create the graph.

```
[18]: def x(t):
          return np.cos(t) + 1/2*np.cos(7*t) + 1/3*np.sin(17*t)
```

```
[19]: def y(t):
          return np.sin(t) + 1/2*np.sin(7*t) + 1/3*np.cos(17*t)
```

```
[20]: x_axis= x(np.linspace(0, 14*np.pi, 1000))
      y_axis= y(np.linspace(0, 14*np.pi, 1000))
```

```
[21]: plt.figure(figsize=(4, 4));
      plt.axis('off');
      plt.plot(x_axis, y_axis, linewidth=0.3);
```

[21]:

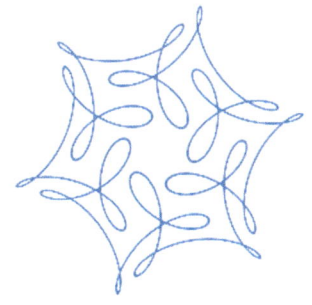

We remark that it is not necessary to define the functions x(t) and y(t). It is enough to define t = np.linspace(0, 14*np.pi, 1000)) and then set

x_axis = np.cos(t) + 1/2*np.cos(7*t) + 1/3*np.sin(17*t)

y_axis = np.sin(t) + 1/2*np.sin(7*t) + 1/3*np.cos(17*t)

Exercise 8.4 *Plot the graph of the function*

$$f(x) = \begin{cases} -x, & \text{if } |x| < 1 \\ \sin(x), & \text{if } 1 \le |x| < 2 \\ \cos(x), & \text{otherwise.} \end{cases}$$

Solution

Given the type of conditions introduced in the body of this function, we revert to defining the function element-wise.

```
[22]: def f(x):
          if abs(x) < 1:
              f = -x
          elif 1 <= abs(x) < 2:
              f = math.sin(x)
          else:
              f = math.cos(x)
          return f
```

```
[23]: import math
      import numpy as np

      x = np.arange(-4, 4, 0.1)
      y = list(map(f, x))
```

The function $f(x)$ is not a continuous function, as becomes more clear from its graph.

```
[24]: plt.plot(x, y, 'r.');
```

[24]: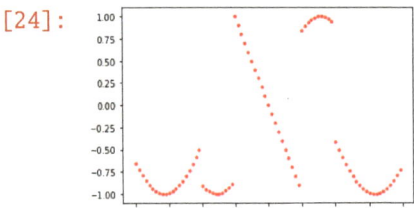

8.2 Plots as Objects

So far we have been using `plot` as a function, giving two list of `x` and `y` for the coordinates to create the graph. However, since `plots` are objects, we can create new plots simply by changing the attributes of an existing `plot`. The following examples show how much more we can get when we work in this manner. We will be using the same data generated for the butterfly graph from the above example.

```
[25]:  fig, ax = plt.subplots()
       plt.figure(figsize=(4, 4),dpi=400);
       ax.axis('equal')
       ax.plot(x_axis, y_axis);
```

[25]: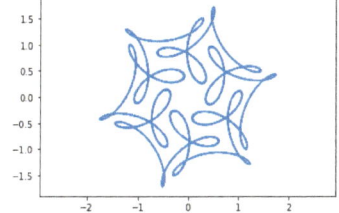

```
[26]:  fig
```

[26]:

Here we have defined the plot object `fig` which comes with is own "Axes" object `ax`. The object `ax` is where we assign the data and decorate the graph, such as adding titles and scales. We can then plot the whole graph using the object `fig`.

A figure object can have several Axes, which then allows us to handle several plots in one go. The line `fig, ax = plt.subplots(m,n)` creates an object `fig`, with $m \times n$ Axes. Each of these Axes can have their own x-y data and specifications, thus the object `fig` contains $m \times n$ graphs. Think of `fig` as an $m \times n$ matrix, with each entry a graph with its own specification. We can access each of these subplots with `ax[i,1]`, again similar to accessing entries of a matrix. The following examples shows how wonderful this approach is.

```
[27]:  def x(t):
           return np.cos(t) + 1/2*np.cos(7*t) + 1/3*np.sin(17*t)

       def y(t):
           return np.sin(t) + 1/2*np.sin(7*t) + 1/3*np.cos(17*t)

       def rrange(i):
```

```
      return np.linspace(0, i * np.pi, 40000)

fig, ax = plt.subplots(3)
ax[0].axis('equal')
ax[1].axis('equal')
ax[2].axis('equal')
ax[0].plot(x(rrange(0.1)), y(rrange(0.1)));
ax[1].plot(x(rrange(1.1)), y(rrange(1.1)));
ax[2].plot(x(rrange(2.1)), y(rrange(2.1)));
```

[27]: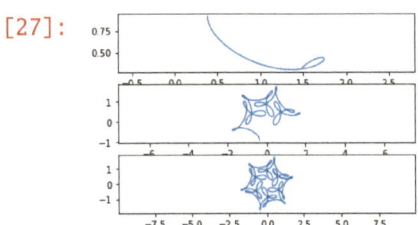

[28]:
```
fig, ax = plt.subplots(6)

for i in [0, 1, 2, 3, 4, 5]:
    ax[i].axis('equal')
    ax[i].plot(x(rrange(0.3 * i )), y(rrange(0.3 * i)))
    ax[i].set_xlim([1, 5])
plt.show()
```

[28]: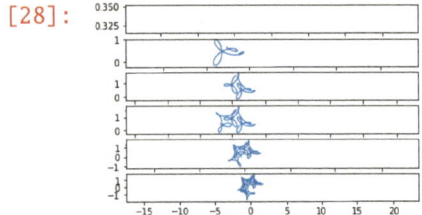

Exercise 8.5 *Define the Conway recursive sequence* $a(1) = 1, a(2) = 1$ *and*

$$a(n) = a(a(n-1)) + a(n - a(n-1))$$

and plot $a(n)/n$, *as n runs from 1 to 1500.*

Next consider a modified version of the function,

$$a_{i,k}(n) = a(a(n-i)) + a(n - a(n-k))$$

for $1 \le i, k \le 2$ *and plot and plot* $a_{i,k}(n)/n$.

Solution

Note that the Conway function $a(n) = a(a(n-1)) + a(n - a(n-1))$ is quite a complicated recursive function; it calls itself twice within its definition. However if we just translate it straight into Python, the program takes care of this recursive behaviour.

```
[29]: def a(n):
          a = [0, 1, 1]
          if n == 2:
              return a[1 : ]
          else:
              for i in range(3, n + 1):
                  a.append(a[a[-1]] + a[i - a[-1]])
              return a[1 : ]
```

```
[30]: import numpy as np
      import matplotlib.pyplot as plt

      y = np.array(a(1500)) / np.arange(1, 1501)
```

```
[31]: plt.plot(y);
```

[31]:

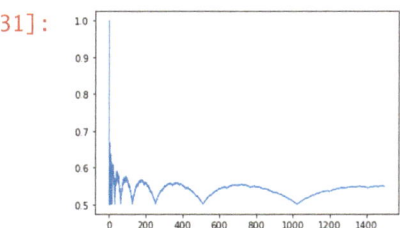

Next we define the modified version of the Conway function.

```
[32]: def a(n, l, k):
          a = [0, 1, 1]
          if n == 2:
              return a[1 : ]
          else:
              for i in range(3, n + 1):
                  a.append(a[a[-l]] + a[i - a[-k]])
              return a[1 : ]
```

```
[33]: fig, ax = plt.subplots(2, 2)

      for l in range(1,3):
          for k in range(1,3):
```

```
            y = np.array(a(1500, 1, k))/np.arange(1, 1501)
            ax[l - 1, k - 1].plot(y)
            ax[l - 1, k - 1].set_title(f'l={l}, k={k}')

for axs in ax.flat:
    axs.label_outer()
```

[33]: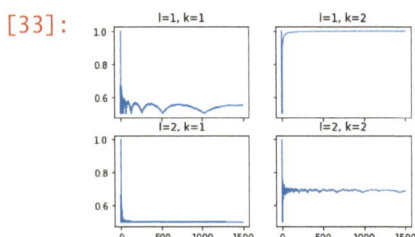

The last loop will hide the x labels and tick labels for the top plots and the y ticks for the right plots.

It is interesting to see that a slight change in the definition of the Conway recursive function makes the behaviour of the sequence quite chaotic. This has been studied by K. Pinn in the article "A chaotic cousin of Conway's recursive sequence", available on arXiv.org.

8.3 Animation

In order to create an animation, we need to enable the interactive plot. This can be done by the following command in Jupyter.

[34]: ```
%matplotlib notebook
```

Next we will use FuncAnimation within the command animation. The idea is to generate several frames of plots and then show them one after the other. The command FuncAnimation just does that. In the line FuncAnimation(fig, animate, frames=100, interval=20, ...) we create a fig, which is our graph object. This, as before, we do with fig, ax = plt.subplots(). We know ax.plot(x,y) will create the graph. The idea is to modify the x and y coordinates accordingly and then ask Python to show them frame by frame with some interval. This is done by defining a function animate which takes care of these changes and frames to send the parameters into animate. The interval sets the delay (in milliseconds) between frames. As usual, we start with examples.
```

```
[35]:  import numpy as np
       import matplotlib.pyplot as plt
       import matplotlib.animation as animation

       fig, ax = plt.subplots()

       ax.set_xlim(0, 10*np.pi)
       ax.set_ylim(-1.1, 1.1)

       line, = ax.plot([], 'r.')

       def animate(i):
           x = np.arange(0, (i/10)*np.pi, 0.1)
           y = np.sin(x)
           line.set_data((x, y))
           return line,

       ani = animation.FuncAnimation(
           fig, animate, frames=100, interval=20, blit=True)
       plt.show()
```

[35]:

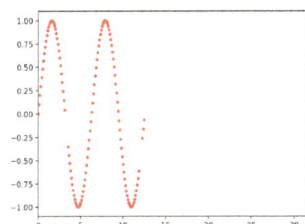

Now that we have the blueprint for creating graphs, we modify the above example to
see how the graph of the parametric equations

$$x(t) = \cos(t) + 1/2\cos(7t) + 1/3\sin(17t)$$
$$y(t) = \sin(t) + 1/2\sin(7t) + 1/3\cos(17t)$$

is created.

```
[36]:  import numpy as np
       import matplotlib.pyplot as plt
       import matplotlib.animation as animation

       def x(t):
           return np.cos(t) + 1/2*np.cos(7*t) + 1/3*np.sin(17*t)
```

```python
def y(t):
    return np.sin(t) + 1/2*np.sin(7*t) + 1/3*np.cos(17*t)

fig, ax = plt.subplots()

ax.set_xlim(-np.pi, np.pi)
ax.set_ylim(-2, 2)

line, = ax.plot([])

def animate(i):
    t = np.linspace(0, (i/10)*np.pi, 40000)
    xc = x(t)
    yc = y(t)
    line.set_data((xc, yc))
    return line,

ani = animation.FuncAnimation(
    fig, animate, frames=100, interval=200, blit=True)

plt.show()
```

[36]:

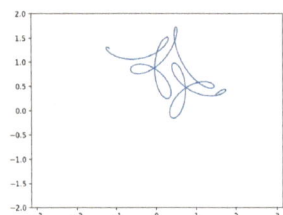

Exercise 8.6 *Plot the polar graph of* $r = 3\cos(6\theta)$.

Solution

For plotting a polar graph, one is given the angle θ and the length $r = 3\cos(6\theta)$. Using trigonometry, we can obtain the (x, y) coordinates as follows:

$$x = r\sin(\theta)$$
$$y = r\cos(\theta)$$

```
[37]:  theta = np.linspace(0, 2 * np.pi, 4000)
       r = 3 * np.cos(6 * theta)
       x = r * np.sin(theta)
       y = r * np.cos(theta)
```

```
[38]:  plt.axis('off');
       plt.axis('equal');
       plt.plot(x,y);
```

[38]:

This is another example of multiple graphs. We are going to plot the polar graph above and then introduce some "noise" to the result using the command random.

```
[39]:  import random
       print(random.random())
```

[39]: 0.43010950529070247

```
[40]:  theta = np.linspace(0, 2 * np.pi, 500)
       r = 3 * np.cos(6 * theta)
       x = r * np.sin(theta)
       y = r * np.cos(theta)

       ran1 = [random.random() for i in range(500)]
       ran2 = [random.random() for i in range(500)]

       fig, ((ax1, ax2), (ax3, ax4)) = plt.subplots(2, 2)
       ax1.axis('equal')
       ax2.axis('equal')
       ax3.axis('equal')
       ax4.axis('equal')
       ax1.plot(x, y);
       ax2.plot(x, y + ran1, color = 'green')
       ax3.plot(x, y + ran2, color = "red")
       ax4.plot(x + ran1, y + ran2, color = 'yellow');
```

[40]: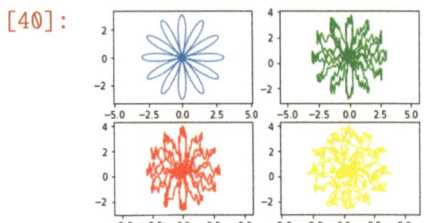

The library matplotlib allows one to create the contour plot of functions as well. The following two examples demonstrate this ability. For this we need to use numpy and the function meshgrid.

Exercise 8.7 *Plot the contour of* $\sin(x^2)\cos(y^2)$ *between* $-\pi$ *and* π.

Solution

Recall the function meshgrid discussed in Chapter 7. Suppose $x = (x_1, x_2, \cdots, x_m)$ and $y = (y_1, y_2, \cdots, y_n)$. Then np.meshgrid(x, y) gives two matrices of the form

```
[[array([[x_1, x_2, ..., x_m],
         [x_1, x_2, ..., x_m],
         [x_1, x_2, ..., x_m],
         ................ ,
         [x_1, x_2, ..., x_m]),
   array([[y_1, y_1, ..., ., y_1],
          [y_2, y_2, ..., ., y_2],
          [. , . , ..., ., . ],
          [. , . , ..., ., . ],
          [y_n, y_n, ., ., y_n]])]
```

We can use meshgrid effectively to create the contours of functions.

[41]:
```python
import numpy as np
import matplotlib.pyplot as plt

points = np.arange(-np.pi, np.pi, 0.01)
xs, ys = np.meshgrid(points, points)
z = np.sin(xs**2) * np.cos(ys**2)
```

[42]:
```python
h = plt.contourf(xs, ys, z)

plt.axis('scaled')
plt.colorbar()
```

```
plt.title('countourplot of $\sin(x^2)   \cos(y^2)$')

plt.show()
```

[42]:

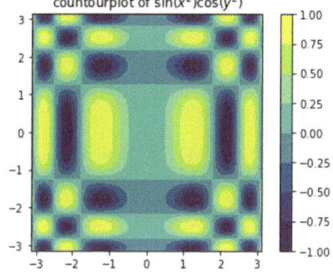

Exercise 8.8 *Consider the function*

$$f(x, y) = \frac{\sin(x^2 + y^2)}{x + y}.$$

Plot the contour of

$$f\Big(f\big(f(x, y), f(x, y)\big), f\big(f(x, y), f(x, y)\big)\Big).$$

Solution

Here again we encounter a complex composition of functions. However, once the function $f(x, y)$ is defined, Python can handle the composition itself. This is yet another example of how to use the function meshgrid to produce contours.

[43]:
```
def f(x, y):
    return np.sin(x**2 + y**2)/(x + y)
```

[44]:
```
import numpy as np
import matplotlib.pyplot as plt

points = np.arange(-np.pi, np.pi, 0.001)
xs, ys = np.meshgrid(points, points)
z = f(xs, ys)
zc = f(f(z , z), f(z, z))
```

```
[45]: h = plt.contourf(xs, ys, zc)

      plt.axis('scaled')
      plt.colorbar()
      plt.title('countourplot')

      plt.show()
```

[45]: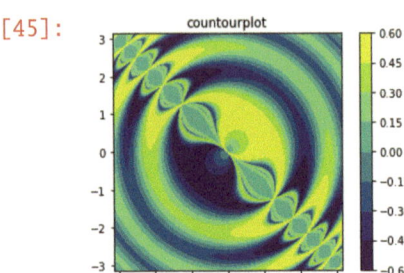

8.4 Case Study, Persian Carpets

The following amusing algorithm was suggested in a short paper by Anne Burns to create "Persian rug-like patterns" using a recursive procedure. (Anne M. Burns, *Persian recursion, Mathematics Magazine*, Vol. 70, No. 3, 1997, 196–199.) Following Burns' paper, we describe the procedure here and then we will write a Python code to generate some of these patterns.

Consider a $2^n + 1 \times 2^n + 1$ celled square. The aim is to colour each cell based on the colour of the corners already coloured. This is achieved by allocating a number between 0 and $m - 1$ to each cell, m being the number of available colours. We begin to colour the outer cells first, colouring them all the same colour in order to form a border. We then apply the following procedure recursively:

• Use the four cells in the corners and a four-variable function to determine a different colour. For example the function could be

$$f(c_1, c_2, c_3, c_4) = (c_1 + c_2 + c_3 + c_4) \quad \mod m.$$

• Allocate the newly determined colour to the interior row cells and column cells in the centre.

• Apply this scheme to all of the four new bordered squares generated at each execution.

Once all cells have been coloured the function will terminate.

Solution

We first translate the procedure into Python, writing a code to generate a matrix with the colours assigned to the entries. This requires a recursive function, finding the middle row and column and then calling the function again four times for each of the four new matrices after the partition.

In order to put the code together, we start with an example. Consider the $2^3 + 1 \times 2^3 + 1$ cells with $m = 7$ and the boundary cells all 4.

$$\begin{pmatrix} 4\,4\,4\,4\,4\,4\,4\,4\,4 \\ 4\,0\,0\,0\,0\,0\,0\,0\,4 \\ 4\,0\,0\,0\,0\,0\,0\,0\,4 \\ 4\,0\,0\,0\,0\,0\,0\,0\,4 \\ 4\,0\,0\,0\,0\,0\,0\,0\,4 \\ 4\,0\,0\,0\,0\,0\,0\,0\,4 \\ 4\,0\,0\,0\,0\,0\,0\,0\,4 \\ 4\,0\,0\,0\,0\,0\,0\,0\,4 \\ 4\,4\,4\,4\,4\,4\,4\,4\,4 \end{pmatrix}.$$

In Python, we consider the matrix x and eventually fill all the entries with the specific numbers.

```
[46]:  import numpy as np
       import matplotlib.pyplot as plt
```

```
[47]:  s = 2**3 + 1
       m = 7
```

```
[48]:  x = np.zeros([s, s])
```

At the moment, we have created an 8×8-matrix with all entries 0. We eventually replace the entries with values coming from the Persian recursion. Note that we could have also used np.empty([s, s]) to generate the generic matrix.

```
[49]:  x
```

```
[49]:  array([[0., 0., 0., 0., 0., 0., 0., 0., 0.],
              [0., 0., 0., 0., 0., 0., 0., 0., 0.],
              [0., 0., 0., 0., 0., 0., 0., 0., 0.],
              [0., 0., 0., 0., 0., 0., 0., 0., 0.],
              [0., 0., 0., 0., 0., 0., 0., 0., 0.],
              [0., 0., 0., 0., 0., 0., 0., 0., 0.],
              [0., 0., 0., 0., 0., 0., 0., 0., 0.],
              [0., 0., 0., 0., 0., 0., 0., 0., 0.],
              [0., 0., 0., 0., 0., 0., 0., 0., 0.]])
```

First we assign the number 4 to the boundaries.

```
[50]:  x[ : ,0] = x[ : ,-1] = x[0, : ] = x[-1, : ] = 4
```

```
[51]:  x
```

```
[51]:  array([[4., 4., 4., 4., 4., 4., 4., 4., 4.],
               [4., 0., 0., 0., 0., 0., 0., 0., 4.],
               [4., 0., 0., 0., 0., 0., 0., 0., 4.],
               [4., 0., 0., 0., 0., 0., 0., 0., 4.],
               [4., 0., 0., 0., 0., 0., 0., 0., 4.],
               [4., 0., 0., 0., 0., 0., 0., 0., 4.],
               [4., 0., 0., 0., 0., 0., 0., 0., 4.],
               [4., 0., 0., 0., 0., 0., 0., 0., 4.],
               [4., 4., 4., 4., 4., 4., 4., 4., 4.]])
```

Now by following the Persian procedure, we get to the numbers in the corners (here all 4) and compute

$$4 + 4 + 4 + 4 \quad \mod 7 = 2$$

which we place in centre of matrix as shown:

$$\begin{pmatrix} 4 & 4 & 4 & 4 & 4 & 4 & 4 & 4 & 4 \\ 4 & \square & \square & \square & 2 & \square & \square & \square & 4 \\ 4 & \square & \square & \square & 2 & \square & \square & \square & 4 \\ 4 & \square & \square & \square & 2 & \square & \square & \square & 4 \\ 4 & 2 & 2 & 2 & 2 & 2 & 2 & 2 & 4 \\ 4 & \square & \square & \square & 2 & \square & \square & \square & 4 \\ 4 & \square & \square & \square & 2 & \square & \square & \square & 4 \\ 4 & \square & \square & \square & 2 & \square & \square & \square & 4 \\ 4 & 4 & 4 & 4 & 4 & 4 & 4 & 4 & 4 \end{pmatrix}$$

In Python we can achieve this by defining the function $cx(l, r, t, b)$, where l, r, t and b are the left, right, top and bottom coordinates of the matrix.

```
[52]:  x.shape
```

```
[52]:  (9, 9)
```

```
[53]:  def cx(l, r, t, b):
           new_col = (x[t,l] + x[t,r] +  x[b,l] + x[b,r]) % m
           return new_col.astype(int)

       l = 0; r = x.shape[0] - 1; t = 0; b = x.shape[1] - 1

       cx(l, r, t, b).astype(int)
```

[53]: 2

With `mc = (l + r)/2`, `mr = (t + b)/2` we find the middle rows and column
and `cx` gives the new colour. This can be done by

```
[54]:  mc = int((l + r)/2); mr = int((t + b)/2)

       x[(t+1) : b, mc] = cx(l, r, t, b)
       x[mr,(l+1) : r] = cx(l, r, t, b)
```

Notice how `x[(t+1) : b, mc]` and `x[mr,(l+1) : r]` now assign the new numbers to the middle columns and rows.

```
[55]:  x
```

```
[55]:  array([[4., 4., 4., 4., 4., 4., 4., 4., 4.],
              [4., 0., 0., 0., 2., 0., 0., 0., 4.],
              [4., 0., 0., 0., 2., 0., 0., 0., 4.],
              [4., 0., 0., 0., 2., 0., 0., 0., 4.],
              [4., 2., 2., 2., 2., 2., 2., 2., 4.],
              [4., 0., 0., 0., 2., 0., 0., 0., 4.],
              [4., 0., 0., 0., 2., 0., 0., 0., 4.],
              [4., 0., 0., 0., 2., 0., 0., 0., 4.],
              [4., 4., 4., 4., 4., 4., 4., 4., 4.]])
```

Next the function `colorgrid[l, r, t, b]` will assign the new numbers to the
entries of the matrix.

```
[56]:  def colorgrid(l, r, t, b):
           if (l < r -1):
               mc = int((l+r)/2); mr = int((t+b)/2)

               x[(t+1):b,mc] = cx(l, r, t, b)
               x[mr,(l+1):r] = cx(l, r, t, b)

               colorgrid(l, mc, t, mr)      # Top left
               colorgrid(mc, r, t, mr)      # Top right
               colorgrid(l, mc, mr, b)      # Bottom left
               colorgrid(mc, r, mr, b)      # Bottom right

       colorgrid(0, x.shape[0]-1, 0, x.shape[1]-1)
       x.astype(int)
       x
```

```
[56]: array([[4., 4., 4., 4., 4., 4., 4., 4., 4.],
             [4., 5., 0., 3., 2., 3., 0., 5., 4.],
             [4., 0., 0., 0., 2., 0., 0., 0., 4.],
             [4., 3., 0., 6., 2., 6., 0., 3., 4.],
             [4., 2., 2., 2., 2., 2., 2., 2., 4.],
             [4., 3., 0., 6., 2., 6., 0., 3., 4.],
             [4., 0., 0., 0., 2., 0., 0., 0., 4.],
             [4., 5., 0., 3., 2., 3., 0., 5., 4.],
             [4., 4., 4., 4., 4., 4., 4., 4., 4.]])
```

Note that, within `colorgrid`, we have called the function four times to take care of the four squares created after partitioning the matrix.

Next we assign colours to the numbers and create our Persian carpet. This can be done by the method `matshow` within `matplotlib`.

```
[57]: mat = np.array([[1,2,3], [4, 5, 6], [7, 8, 9]])
      plt.matshow(mat);
```

[57]:

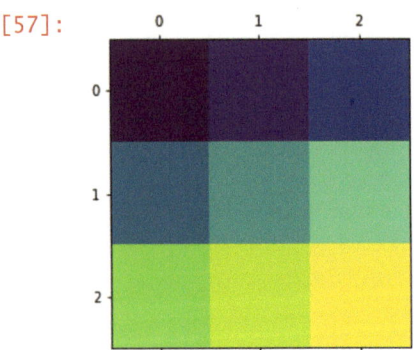

With our example, we get the following carpet.

```
[58]: colorgrid(0, x.shape[0]-1, 0, x.shape[1]-1)
      x.astype(int)
      plt.matshow(x);
```

[58]: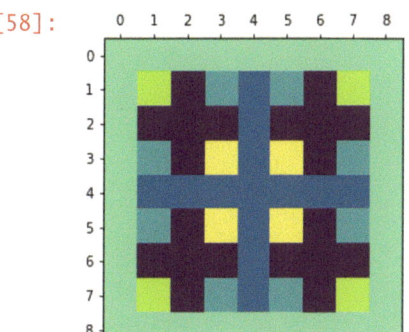

Putting all these together, we define the function perCarpet to generate interesting patterns.

```
[59]: def perCarpet(m, n, z):

          x = np.empty([2**n+1, 2**n+1])
          x[ : ,0] = x[ : ,-1] = x[0, : ] = x[-1, : ] = z

          def cx(l, r, t, b):
              new_col = (x[t,l] + x[t,r] +  x[b,l] + x[b,r]) % m
              return new_col.astype(int)

          def colorgrid(l, r, t, b):
              if (l < r - 1):
                  mc = int((l + r)/2); mr = int((t + b)/2)

                  x[(t + 1):b, mc] = cx(l, r, t, b)
                  x[mr,(l + 1): r] = cx(l, r, t, b)

                  colorgrid(l, mc, t, mr)   # Top left
                  colorgrid(mc, r, t, mr)    # Top right
                  colorgrid(l, mc, mr, b)    # Bottom left
                  colorgrid(mc, r, mr, b)    # Bottom right

          colorgrid(0, x.shape[0]-1, 0, x.shape[1]-1)
          x.astype(int)
          return plt.matshow(x)
```

```
[60]: perCarpet(7, 3, 4);
```

[60]: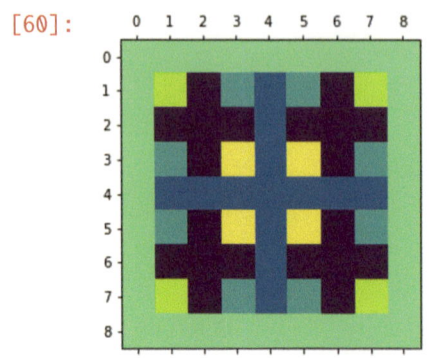

[61]:
```
perCarpet(9, 10, 1);
```

[61]: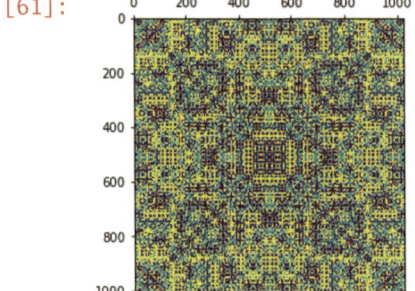

[62]:
```
perCarpet(5, 8, 7);
```

[62]: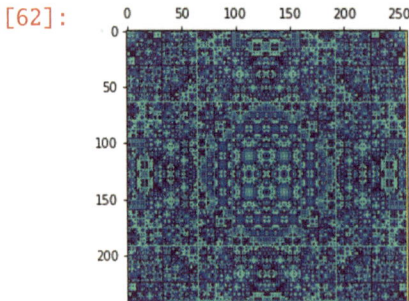

8.5 Case Study, Predictive Estimation

In many applications as the size increases, the time it takes to complete the task also increases. The following case study is a good example showcasing how we can analyse the time dependence of an algorithm. Let us donate the size of an application by **n** and the time it takes to run the program and obtain the result by **t**. It rarely happens that the relationship between the size and time is linear, i.e., $\mathbf{t} = k\,\mathbf{n} + l$, for some constants k and l. If this is the case, then by sampling some small **n** and measuring **t**, and plotting $(\mathbf{n}, \mathbf{t})$, we should obtain a straight line. If this is not the case (and again oftentimes it is not), and if the relation is of the form

$$\mathbf{t} = k\,\mathbf{n}^{l}$$

or

$$\mathbf{t} = k\,l^{\mathbf{n}},$$

then we might also be able to find the relation by running the program for small values of **n**.

Suppose the relation is of the form $\mathbf{t} = k\,\mathbf{n}^{l}$. Then taking logs we have

$$\log(\mathbf{t}) = l\log(\mathbf{n}) + \log(k).$$

Now if we run the application for some (small) values of **n**, measure the time it takes to complete the task, and then plot $\log(\mathbf{t})$ versus $\log(\mathbf{n})$, we should end up with a straight line. This will then be an indication that the relation between **t** and **n** is indeed of the form $\mathbf{t} = k\,\mathbf{n}^{l}$. If the graph we obtain does not look like a straight line, then we can consider the next case of $\mathbf{t} = k\,l^{\mathbf{n}}$. Taking logs we get

$$\log(\mathbf{t}) = \mathbf{n}\log(l) + \log(k).$$

In this case, the graph of $\log(\mathbf{t})$ versus **n** should create a straight line.

Once this is done, we can use regression methods available in Python to get a least-squares best fit linear approximation to the line. Thus we get an equation of line $y = ax + b$ which approximates the sample. Once we have the equation of the line, we should be able to calculate the constants k and l. This is done by using the identity $e^{\log(x)} = x$.

Case study

Let B_n denote the $n \times n$ matrix with (i, j)-th entry equal to

$$b_{ij} = \begin{cases} \frac{1}{i^2 - j} & \text{if } i > j \\ \frac{1}{j - i} & \text{if } j > i \\ 0 & \text{if } i = j. \end{cases}$$

Look at the numerical values of the determinant of B_n for $3 \le n \le 30$ and display these values graphically. You should observe that the values seem to follow a pattern. Look at the sequence and show that the n-Log or Log-Log transformation gives a fit *that is approximately linear and hence obtain and test a formula that predicts the value of the sequence. Define a function that predicts the value of the determinant of B_n. What is the largest percentage error that your formula has for $30 \leq n \leq 50$?*

Solution

We start by importing the libraries needed and then define the matrix.

```python
[63]: import numpy as np
      import matplotlib.pyplot as plt
```

```python
[64]: def B(n):
          mat = np.empty([n, n])
          for i in range(n):
              for j in range(n):
                  i += 1; j += 1
                  if (i > j):
                      e = 1/(i**2 - j)
                  elif (j > i):
                      e = 1/(j - i)
                  else:
                      e = 0
                  i -= 1; j -= 1
                  mat[i, j] = e
          return mat

      B(3)
```

```
[64]: array([[0.        , 1.        , 0.5       ],
             [0.33333333, 0.        , 1.        ],
             [0.125     , 0.14285714, 0.        ]])
```

```python
[65]: B(5)
```

```
[65]: array([[0., 1., 0.5, 0.33333333, 0.25],
             [0.33333333, 0., 1., 0.5, 0.33333333],
             [0.125, 0.14285714, 0., 1., 0.5],
             [0.06666667, 0.07142857, 0.07692308, 0., 1.],
             [0.04166667, 0.04347826, 0.04545455, 0.04761905, 0.]])
```

The library numpy allows us to evaluate the determinant of a matrix. We compute the determinant of $B(n)$ for certain values of n.

```python
[66]: det_B = [np.linalg.det(B(n)) for n in range(2, 31)]
```

We look at a sample of the determinants.

```
[67]: [det_B[i].round(5) for i in range(8)]
```

```
[67]: [-0.33333, 0.14881, -0.05891, 0.03163, -0.02061, 0.01468,
       -0.01104, 0.00861]
```

Note that, it looks like $\det(B(n))$ is positive for even values of n and negative for odd values. It is now easy to plot the values and observe the pattern.

```
[68]: x = range(2, 31)
```

```
[69]: plt.ylim(-0.015, 0.015)
      plt.plot(x, det_B, 'r^');
```

[69]:

As one can see there are two sub-sequences here. We treat them separately.

```
[70]: det_B[ : : 2]
```

```
[70]: [-0.3333333333333333,
       -0.05890949328449329,
       -0.020606972438086804,
       -0.011037045008369608,
       -0.00691202302161902,
       -0.004740845068644299,
       -0.0034551244433575485,
       -0.0026303788255851936,
       -0.002069586219401722,
       -0.0016709092704200235,
       -0.0013773309712474416,
       -0.0011548848970943254,
       -0.0009823045823221005,
       -0.0008457200609293778,
       -0.0007357709130155691]
```

```
[71]: det_B[1 : : 2]
```

```
[71]:   [0.1488095238095238,
         0.03162805947646787,
         0.01468496294836567,
         0.008611643884091177,
         0.005672896223774175,
         0.0040217496236219985,
         0.0030006019535249876,
         0.0023247863447106655,
         0.0018542623250821055,
         0.0015134903892876795,
         0.001258765078394007,
         0.001063359924241922,
         0.0009101799734072838,
         0.0007878755713934514]
```

```
[72]:   plt.ylim(-0.015, 0.015)
        plt.plot(x[ : : 2], det_B[ : : 2], 'r^');
```

[72]:

```
[73]:   plt.ylim(-0.015, 0.015)
        plt.plot(x[1 : : 2], det_B[1 : : 2], 'g^');
```

[73]:

```
[74]:   X = np.array(x[1 : : 2])
        Y = np.array(det_B[1 : : 2])
```

```
[75]:   X
```

```
[75]:   array([ 3,   5,   7,   9,  11,  13,  15,  17,  19,  21,  23,  25,  27,
               29])
```

```
[76]: Y
```

```
[76]: array([0.14880952, 0.03162806, 0.01468496, 0.00861164,
             0.0056729,  0.00402175, 0.0030006 , 0.00232479,
             0.00185426, 0.00151349, 0.00125877, 0.00106336,
             0.00091018, 0.00078788])
```

Now we have the sample values for n (in the list **X**) and t (in the list **Y**). Next we need to determine whether the relationship between n and t is of the form $n - \log$ or log-log. First we plot $(\mathbf{n}, \log(\mathbf{t}))$ for the samples we have.

```
[77]: plt.plot(X, np.log(Y), 'g^');
```

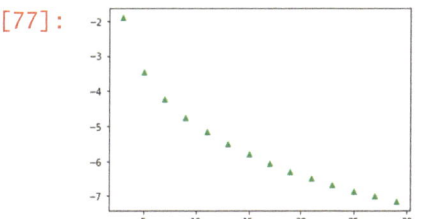

This clearly shows the relationship between **n** and **t** is not of the form $\mathbf{t} = k\, l^{\mathbf{n}}$. We now plot $(\log(\mathbf{n}), \log(\mathbf{t}))$.

```
[78]: plt.plot(np.log(X), np.log(Y), 'g^');
```

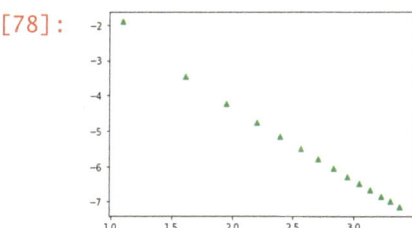

This clearly shows the relation is of the form $\mathbf{t} = k\, \mathbf{n}^l$. We now use the regression facilities in Python to obtain the approximation. We need to import `LinearRegression`. The rest of the code is self-explanatory.

```
[79]: X = np.log(X); Y = np.log(Y)
```

```
[80]: from sklearn.linear_model import LinearRegression
```

```
[81]: X = X.reshape((-1, 1))
```

```
[82]: X
```

```
[82]: array([[1.09861229],
             [1.60943791],
             [1.94591015],
             [2.19722458],
             [2.39789527],
             [2.56494936],
             [2.7080502 ],
             [2.83321334],
             [2.94443898],
             [3.04452244],
             [3.13549422],
             [3.21887582],
             [3.29583687],
             [3.36729583]])
```

```
[83]: model = LinearRegression()
      model.fit(X, Y)
```

```
[83]: LinearRegression()
```

We could have done this in the following way as well.

```
[84]: model = LinearRegression().fit(X, Y)
```

We can now ask the model that we devised to predict the values of the (log of) determinants and compare them with the actual values.

```
[85]: y_pred = model.predict(X)
```

```
[86]: y_pred
```

```
[86]: array([-2.20422019, -3.33615459, -4.08174073, -4.63862635,
             -5.08329094, -5.45346474, -5.77056076, -6.04790876,
             -6.29437275, -6.51614689, -6.71773052, -6.90249516,
             -7.07303251, -7.23137786])
```

```
[87]: plt.plot(X, Y, 'g^');
      plt.plot(X, y_pred, color='blue')
      plt.show()
```

[87]: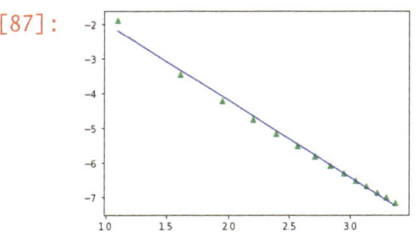

```
[88]: a = model.coef_[0]
      b = model.intercept_

      print(f'The linear regression is y = {a.round(2)} x + {b.
      ↪round(2)}')
```

[88]: The linear regression is y = -2.22 x + 0.23

Since the equation $y = -2.22x + 0.23$ models the equation $\log(\mathbf{t}) = l\log(\mathbf{n}) + \log(k)$, in order to obtain $\mathbf{t}$, i.e., the actual determinants, we use the identity $e^{\log(\mathbf{t})} = \mathbf{t}$.

```
[89]: def det_predic(n):
          return np.exp(a * np.log(n) + b)
```

```
[90]: print('\tActual determinant\tPrediction')
      for i in x[1 : : 2]:
          print('  ', det_B[i - 2], '              ', det_predic(i))
```

[90]:   Actual determinant   Prediction
      0.1488095238095238   0.11033653319196238
      0.03162805947646787   0.03557348953553673
      0.01468496294836567   0.01687805999608341
      0.008611643884091177   0.009670973020585037
      0.005672896223774175   0.006199473332366782
      0.0040217496236219985   0.004281444854281032
      0.0030006019535249876   0.0031180085833193575
      0.0023247863447106655   0.0023627980204066015
      0.0018542623250821055   0.0018466672457568213
      0.0015134903892876795   0.0014793582700115937
      0.001258765078394007   0.0012092795372936683
      0.001063359924241922   0.0010052739786632717
      0.0009101799734072838   0.0008476586716945798
      0.0007878755713934514   0.0007235232635160191

```
[91]: print('actual determinant\tprediction')
      for i in range(31, 50, 2):
          print('  ', np.linalg.det(B(i)).round(5), '          ', ⌴
          ↪det_predic(i).round(5))
```

```
[91]: actual determinant        prediction
             0.00069             0.00062
             0.00061             0.00054
             0.00054             0.00048
             0.00048             0.00042
             0.00043             0.00038
             0.00039             0.00034
             0.00036             0.0003
             0.00033             0.00027
             0.0003              0.00025
             0.00027             0.00023
```

```
[92]: def error(n):
          d = np.linalg.det(B(i))
          g = det_predic(i)
          return 100 * ( g - d) / d

      max([error(n) for n in range(31, 50, 2)])
```

```
[92]: -17.390995636384897
```

We have analysed the list of **n** for odd values. A similar treatment can be done for the even values. We leave it as an exercise to the reader.

8.6 Case Study, the Thue–Morse Sequence

To get the Thue–Morse sequence, start with 0 and then repeatedly replace 0 with 01 and 1 with 10. So the first four numbers in the sequence are

$$0$$
$$01$$
$$0110$$
$$01101001$$

Write a function to produce the n-th element of this sequence.

Now generate the 15th number in the Thue–Morse sequence. Then, based on this number, create a graph as follows: starting from the coordinate $(0, 0)$, move ahead by one unit if you encounter 1 in the Thue–Morse sequence and rotate counterclockwise by an angle of $-\pi/3$ if you encounter 0. (The resulting curve converges to the Koch snowflake, a fractal curve of infinite length containing a finite area – see, for example, the Thue–Morse sequence in Wikipedia.)

Solution

We start by creating the Thue–Morse sequence, using the following clever code!

```
[93]:   def ThueMorse(n):
            x = '0'
            for i in range(n-1):
                x = ''.join(['01' if char == '0' else '10' for char
        ⌐in [*x]])
            return x
```

```
[94]:   for i in range(1, 8):
            print(f'{ThueMorse(i)} --the length--⌐
        ⌐{len(ThueMorse(i))}')
```

```
[94]:   0 --the length-- 1
        01 --the length-- 2
        0110 --the length-- 4
        01101001 --the length-- 8
        0110100110010110 --the length-- 16
        01101001100101101001011001101001 --the length-- 32
        0110100110010110100101100110100110010110011010010110100110
        10110 --the length-- 64
```

Note how quickly the sequence grows.

```
[95]:   len(ThueMorse(15))
```

```
[95]:   16384
```

```
[96]:   from math import sin, cos, pi
        import matplotlib.pyplot as plt
        import numpy as np

        directions = ThueMorse(15)
        L = [[0, 0]]
        facing = 0

        for step in [*directions]:
```

```
    if step == "1":
        L.append([L[-1][0] + cos(facing * pi / 3), L[-1][1] +
    sin(facing * pi / 3)])
    else:
        facing = (facing + 1)
x = np.array(L)[:, 0]
y = np.array(L)[:, 1]

plt.figure(figsize=(5, 5))
plt.xlim([np.amin(L) - 1, np.amax(L) + 1])
plt.ylim([np.amin(L) - 1, np.amax(L) + 1])
plt.plot(x, y, color='green')
plt.show()
```

[96]:
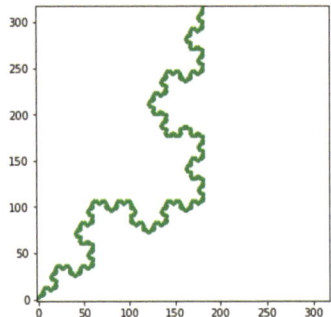

Problems

1. Plot the graph

$$f(x) = \cos^2(x) - e^{-\sum_{k=1}^{30} \frac{\cos(kx)}{1+k}} + e^{-\sum_{k=1}^{50} \frac{\sin(kx)}{1+k}}$$

for x ranging over the interval $[0, 2\pi]$. Then find (numerically) the root of the equation $f(x) = 0$ lying between 4 and 5.

2. Plot the graph of the expression

$$x(2\pi - x) \sum_{n=1}^{50} \frac{\sin(nx)}{n}$$

for $0 \le x \le 2\pi$.

3. Consider the function $\sin(x^2)\cos(y^2)$. Plot the function and its contour. Create an animation, observing how the areas change as the plot develops.

4. Let B_n denote the $n \times n$ matrix with (i, j)-th entry equal to $1/(|i - j|)$ if $i \neq j$ and 0 if $i = j$. Find a simple formula to predict the value of the determinant of B_n for large values of n. What is the largest percentage error that your prediction has for $20 \leq n \leq 40$? (*Hint:* Look at the determinants of B_n for $2 \leq n \leq 10$ first. You will see that predictions are only possible for certain subsequences of the integers. Obtain sufficient data for each of these subsequences so that you can find a transformation that is approximately linear and hence obtain and test a formula for each subsequence. Put the resulting formulae together for the final predictive function.)

5. Consider the first 10000 digits of $\sqrt{2}$ and present them as a "random walk" by converting them into base 4, representing 4 directions (up, down, left and right). We know that $\sqrt{2}$ is an irrational number and irrational numbers have decimal expansions that neither terminate nor become periodic. Write a code to produce this random walk. Try this code with $\sqrt{3}$, $\sqrt{6}$ and $\sqrt{13}$. Is there any comparison one can make among these numbers?

6. (For this problem, see Case study 8.5.) A double sequence $(a_{(m,n)})$ is defined by the rules

$$a_{(1,1)} = a_{(1,2)} = a_{(2,1)} = 1$$
$$a_{(m,1)} = a_{(m-1,1)} + a_{(m-2,1)} \quad (\text{if } m > 2)$$
$$a_{(1,n)} = a_{(1,n-1)} + a_{(1,n-2)} \quad (\text{if } n > 2)$$
$$a_{(m,n)} = a_{(m,1)} + a_{(1,n)} - a_{(m-1,n-1)} \quad (\text{if } m > 1 \text{ and } n > 1).$$

The value of $a_{(10,10)}$ is 67. Produce a table comparing the values of $a_{(n,n)}$ with n for $1 \leq n \leq 20$. Find a simple function that predicts the size of $a_{(n,n)}$ and compare your predictions for $a_{(30,30)}$ and $a_{(100,100)}$ with the correct values.

If the definition of $a_{(m,n)}$ is altered so that

$$a_{(m,n)} = a_{(m,1)} + a_{(1,n)} - 2a_{(m-1,n-1)} \quad (\text{if } m > 1 \text{ and } n > 1)$$

then the terms $a_{(n,n)}$ alternate in sign. By considering separately the sequences $a_{(2,2)}, a_{(4,4)}, \ldots$ and $a_{(1,1)}, a_{(3,3)}, a_{(5,5)}, \ldots$ produce (and verify) a simple formula for estimating the values of $a_{(n,n)}$ in this case.

7. (For this problem, see Case study 8.5.) Define an $n \times n$ matrix $A_n = (a_{ij})$ whose i, j-th entries are

$$a_{ij} = \begin{cases} i + j & \text{if } i = j \\ i^{+j} & \text{if } i \neq j. \end{cases}$$

Find out how long it takes Python to evaluate the inverse of $A(n)$ for $n = 15, 16, \ldots 30$ (this might take some seconds). Hence obtain a formula which estimates how long it takes to calculate the inverse of $A(n)$ in general. Test this formula experimentally by timing the calculation of the inverse of $A(31)$. Use your formula to estimate how long it would take to find the inverse of $A(100)$.

8. Plot the graph of the function

$$f(x) = \begin{cases} -x, & \text{if } |x| < 1 \\ \sum_{n=1}^{10} \sin(\frac{x}{n}), & \text{if } 1 \le |x| < 2 \\ \sum_{n=1}^{10} \cos(\frac{x}{n}), & \text{otherwise.} \end{cases}$$

How many roots does $f(x)$ have in the interval $[-4, 4]$?

Further Reading

There are many excellent books about Python. Once one knows the core of the language, one can easily use these books. Here we collect a few books that are worth consulting.

• Eric Matthes, *Python Crash Course*, 3nd Edition, No Starch Press, 2022.

This comprehensive beginner-level book on Python contains many interesting worked out projects.

• Jake VanderPlas, *A Whirlwind Tour of Python*, O'Reilly Media, 2016.

This excellent book gives a concise overview of how to use Python and the philosophy behind its design.

• Michael T. Goodrich, Roberto Tamassia, and Michael H. Goldwasser, *Data Structures and Algorithms in Python*, Wiley, 2013.

This is an introduction to practical computer science and an excellent source for mathematics and data science students.

• Mark Lutz, *Learning Python*, 5th Edition, O'Reilly Media, 2013.

A comprehensive treatment of Python for more experienced readers which can be used as a reference book as well.

• Eli Stevens, Luca Antiga, and Thomas Viehmann, *Deep Learning with PyTorch*, Manning, 2020.

This book provides more advanced uses of Python in data science.

© The Author(s), under exclusive license to Springer Nature Switzerland AG 2023
R. Hazrat, *A Course in Python*, Springer Undergraduate Mathematics Series,
https://doi.org/10.1007/978-3-031-49780-3

- T. Andreescu, R. Gelca, *Mathematical Olympiad Challenges*, Birkhäuser, 2nd ed. 2009.

- S. Rabinowitz, *Index to Mathematical Problems 1980–1984*, Math pro Press. 1992.

Many of the examples in the present book have been inspired by mathematical problems from previous mathematical olympiads. The above two books have a wonderful collection of such problems.

Index

© The Author(s), under exclusive license to Springer Nature Switzerland AG 2023
R. Hazrat, *A Course in Python*, Springer Undergraduate Mathematics Series,
https://doi.org/10.1007/978-3-031-49780-3